Marine Ecotourism

ASPECTS OF TOURISM

Series Editors: Professor Chris Cooper, *University of Queensland, Australia,*
Dr Michael Hall, *University of Otago, Dunedin, New Zealand*
and Dr Dallen Timothy, *Arizona State University, Tempe, USA*

Aspects of Tourism is an innovative, multifaceted series which will comprise authoritative reference handbooks on global tourism regions, research volumes, texts and monographs. It is designed to provide readers with the latest thinking on tourism world-wide and in so doing will push back the frontiers of tourism knowledge. The series will also introduce a new generation of international tourism authors, writing on leading edge topics.

The volumes will be readable and user-friendly, providing accessible sources for further research. The list will be underpinned by an annual authoritative tourism research volume. Books in the series will be commissioned that probe the relationship between tourism and cognate subject areas such as strategy, development, retailing, sport and environmental studies. The publisher and series editors welcome proposals from writers with projects on these topics.

Other Books in the Series

Classic Reviews in Tourism
Chris Cooper (ed.)
Dynamic Tourism: Journeying with Change
Priscilla Boniface
Journeys into Otherness: The Representation of Differences and Identity in Tourism
Keith Hollinshead and Chuck Burlo (eds)
Natural Area Tourism: Ecology, Impacts and Management
D. Newsome, S.A. Moore and R. Dowling
Progressing Tourism Research
Bill Faulkner, edited by Liz Fredline, Leo Jago and Chris Cooper
Tourism Collaboration and Partnerships
Bill Bramwell and Bernard Lane (eds)
Tourism and Development: Concepts and Issues
Richard Sharpley and David Telfer (eds)
Tourism Employment: Analysis and Planning
Michael Riley, Adele Ladkin, and Edith Szivas
Tourism in Peripheral Areas: Case Studies
Frances Brown and Derek Hall (eds)

Other Books of Interest

Global Ecotoursim Policies and Case Studies
Michael Lück and Torsten Kirstges (eds)

Please contact us for the latest book information:
Channel View Publications, Frankfurt Lodge, Clevedon Hall,
Victoria Road, Clevedon, BS21 7HH, England
http://www.channelviewpublications.com

ASPECTS OF TOURISM 7
Series Editors: Chris Cooper (*University of Queensland, Australia*),
Michael Hall (*University of Otago, New Zealand*)
and Dallen Timothy (*Arizona State University, USA*)

Marine Ecotourism

Issues and Experiences

Edited by
Brian Garrod and Julie C. Wilson

CHANNEL VIEW PUBLICATIONS
Clevedon • Buffalo • Toronto • Sydney

BG – To Lydia, and to the future she will inherit

JCW – For my parents

Library of Congress Cataloging in Publication Data
Marine Ecotourism: Issues and Experiences/Edited by Brian Garrod and Julie C. Wilson.
Aspects of Tourism: 7
Includes bibliographical references.
1. Marine ecotourism. I. Garrod, B. (Brian). II. Wilson, Julie C. III. Series.
G156.5.M36 M37 2005
338.4'791'09162–dc21 2002014382

British Library Cataloguing in Publication Data
A catalogue entry for this book is available from the British Library.

ISBN 1-853150-42-3 (hbk)
ISBN 1-853150-41-5 (pbk)

Channel View Publications
An imprint of Multilingual Matters Ltd

UK: Frankfurt Lodge, Clevedon Hall, Victoria Road, Clevedon BS21 7SJ.
USA: 2250 Military Road, Tonawanda, NY 14150, USA.
Canada: 5201 Dufferin Street, North York, Ontario, Canada M3H 5T8.
Australia: Footprint Books, PO Box 418, Church Point, NSW 2103, Australia.

Typeset by Archetype-IT Ltd (http://www.archetype-it.com).
Printed and bound in Great Britain by the Cromwell Press.

Contents

Foreword

Experts say that whalewatching has become a billion-dollar business around the globe. The migration of the largest animals on the planet is a sight to behold. For travellers, seeing the same species, perhaps even the same animal, is quite a different experience depending on where you go whalewatching.

Marine ecotourism takes on additional importance this year. The United Nations has declared 2002 the 'International Year of Ecotourism'. This ought to be a time for celebration as those practising authentic 'ecotourism' network with other like-minded enterprises. The larger public is becoming increasingly aware of the connections between tourism and the environment and 2002 could be the year when ecotourism achieves its potential around the globe.

While marketing studies can be debated, the value of whales is undeniable to the tourism industry. What was once a small cottage industry has transformed into a global industry. 'A live whale is worth more than a dead whale', Mexican ecotourism consultant Hector Ceballos Lascuraín once told me. How then should we protect this living treasure?

Understanding marine ecotourism is now much more clear thanks to the publication of *Marine Ecotourism: Issues and Experiences*. By providing issues and field experiences, editors Brian Garrod and Julie C. Wilson have done us all a great service. With luck, this volume will educate decision-makers and various stakeholders.

Ron Mader

References

Whalewatching on the Web. On WWW at http://www.planeta.com/ecotravel/resources/whales.html

Marine Tourism Forum. On WWW at http://groups.yahoo.com/group/ballenas-whales

Ron Mader lives in Mexico and travels frequently throughout the Americas. He hosts the award-winning Planeta.com website [http://www.planeta.com] and is the author of the *Mexico: Adventures in Nature* guidebook and the *Exploring Ecotourism in the Americas Resource Guide*.

Preface

Ecotourism is reputed to attract high-spending tourists and estimates suggest it to be 'worth' between US$10 and US$17.5 billion worldwide (Fennell, 1999). As a subset of ecotourism, marine ecotourism is becoming increasingly significant and its economic potential is being increasingly recognised. Beyond the dimension of economic returns, however, it is generally agreed that marine ecotourism has considerable potential to generate a range of wider benefits. Marine ecotourism can, for example, help generate funds for research on marine species and habitats, help fund conservation programmes, assist in raising the profile of marine resources in the planning process, provide an economic rationale for environmental stewardship and provide a focus for the social and cultural regeneration of coastal communities. This having been said, there have been very few proper empirical studies of the actual social and economic impacts of ecotourism in the marine environment (whale-based ecotourism being the principal exception) and even fewer studies of the environmental impacts of marine ecotourism.

While most studies of the economic scale of marine ecotourism have tended to be partial in scope and often lacking in rigour, it is nevertheless clear that marine ecotourism is a major growth area within ecotourism, which is itself a major growth area within the tourism industry as a whole. There has certainly been a widespread realisation that watching whales can be more profitable than commercial whaling (as was argued by Hector Ceballos-Lascuraín). Indeed, it is argued that whale-watching is now more economically valuable than commercial whaling ever was. In recognition of the growing global phenomenon of ecotourism, the year 2002 has been designated the *International Year of Ecotourism* (IYE 2002) by the United Nations Environment Programme (UNEP). At the time of writing, the World Ecotourism Summit is about to take place in Quebec City, Canada, and several regional events have been staged as part of the UN year. The run-up to the year was, however, fraught with controversy, with several major international non-governmental organisations (NGOs) calling for the event to be re-designated as the '*International Year of Reviewing Ecotourism*' (IYRE, 2002). The principal concern has been that the term 'ecotourism' has been widely and readily employed as a marketing buzzword – a tool for latching on to the good intentions of tourists without tour providers having to make any fundamental changes to their products or practices. As academics and practitioners working in the field of ecotourism know only too well, this has often been to the detriment of local environments, economies and host communities. Yet the ideal state of ecotourism is one that presents local communities with a sustainable development alternative – one that benefits local ecosystems, local economies and local people themselves.

Arguably, marine ecotourism development can also contribute to improving the quality of the tourism product – particularly the quality of destination environments. In this respect, however, there is a growing awareness that marine ecotourism should embody the objectives of sustainable development. Experience suggests that unless tourism activity of any kind is properly planned and managed, it risks compromising the economic, social and/or environmental components of the sustainability of an area (Hohl & Tisdell, 1995; Burton, 1998). It may also, because of its transport implications, threaten the global environment. Not only must any ecotourism product be managed to increase the level of sustainability at the local destination level, but the global environmental impact of this form of tourism can be reduced if the energy consumption of the travel element of the holiday can be minimised. The travel element is an unavoidable part of all tourism and the promotion of tourism to peripheral areas – areas that tend to be relatively well endowed with opportunities for marine ecotourism to be developed – will tend automatically to increase proportionately its negative impacts. It is clear, therefore, that the growth of marine ecotourism activity should be a principal concern for policy-makers, practitioners and researchers worldwide, not only in terms of its local sustainability implications but also in terms of its transport-related environmental impacts at the global level.

Why this book was written

Why did we produce this book? Apart from the fact that marine ecotourism is a major growth area (as several of the chapters will argue), one of the main reasons was a response to the growing number of university courses with an ecotourism focus, particularly at postgraduate level. Many universities worldwide have recently developed new Masters Degrees in sustainable tourism, tourism and sustainability (e.g. the University of the West of England, Bristol, UK) and tourism more generally. Others (e.g. the University of Portsmouth, UK) now have Masters Degrees specifically focused on the subject of ecotourism. The area is also becoming very popular as a topic among research postgraduates. This trend is apparent not just in the UK but worldwide, particularly in North America and Australasia.

Another reason is the increasing popularity of ecotourism as a subject of academic research, which can be gauged by the number of new books (e.g. Weaver's *Encyclopedia of Ecotourism*, see Weaver [2001]) and journals (e.g. *Journal of Ecotourism*) that are now in publication. However, the focus has thus far been mainly on ecotourism in general or on ecotourism specifically in terrestrial environments. We felt that this bias required rectification, particularly given the rapid growth in popularity and significant media profile given to whalewatching and scuba-diving, two of the most important candidates for the application of the term 'marine ecotourism'. To the best of our knowledge, this is the first major published book that deals specifically with marine ecotourism.

Finally, this book is a response to the growth of NGOs with a remit in the area of ecotourism (e.g. The International Ecotourism Society, TIES), which is indicative that this area is attracting considerable attention from tourism, environmental and development professionals around the world. The EU Interreg IIc project *Marine Ecotourism for the Atlantic Area* (META-), from which some of the chapters in this book are derived, has also raised a great deal of interest among local government planning departments along the EU Atlantic periphery.

Purpose, aims and focus of this book

The purpose of this book is to inform readers about the concept of marine ecotourism and the key issues involved in ensuring that marine ecotourism is developed in a genuinely sustainable manner. We hope that this book will introduce the reader to the concept and historical development of marine ecotourism, including recent debates questioning the true value of ecotourism as a sustainable development option. The major principles of marine ecotourism, as it ought to be understood as a genuinely sustainable form of tourism, are explained in detail and debated in different local and national contexts.

Meanwhile several chapters in this book explore some major methodological issues concerned with identifying, defining, analysing and evaluating marine ecotourism in different contexts. We have attempted in this volume to highlight the importance of effective community participation and tourist education in ensuring that marine ecotourism lives up to the high ideals it sets itself. Finally, we consider some of the main practical issues associated with marine ecotourism development projects using case studies from around the world.

We hope that by having drawn together some major conclusions relating to the past, present and future development of marine ecotourism, we can stimulate extensive discussion as to what the future might hold for this rapidly expanding phenomenon in terms of policy, practice and research.

With respect to the geographical coverage of the authors and cases featured in this book, the selection is very international in outlook. We hope that the perspectives of contributors living and/or working in the UK, Ireland, Spain, Uruguay, Canada, New Zealand, Australia, Greece and India will reinforce the usefulness of this book in an international context. Similarly, case studies of experiences with marine ecotourism are provided from Europe, the Asia-Pacific region, North America and Latin America.

How this book was written

Some of the chapters in this book were originally presented as papers at an international workshop on marine ecotourism: a parallel session of the European Association of Leisure and Tourism Education (ATLAS) conference, held in Dublin,

Ireland, in October 2001. However, in the interests of developing the book into a coherent volume, rather than a volume of conference proceedings, the contributors were asked to re-work their chapters in the light of some of the things they had learned at the international workshop. Chapter authors were then encouraged to make connections between those ideas and the material presented in their own chapters. In addition, to ensure that the book became much more than a collection of conference papers, several new chapters were invited from authors renowned in the field of marine tourism. This process not only expanded the geographical scope of the perspectives and specific case studies but also resulted in a more complete volume, covering even more elements of marine ecotourism as an entire topic area. An introduction to the book was written in order to provide some context to marine ecotourism and to give the book shape. Meanwhile, a concluding section was prepared in an attempt to synthesise some of the major points of agreement and dissension among the contributors in respect of the major issues that are considered to face marine ecotourism at present.

Some final thoughts

We believe that there is a unique focus to this book, in that it treats marine ecotourism as a subject that is in many important respects very different to terrestrial ecotourism. This prospect, along with the global interest that has been generated in the run-up to and during the IYE 2002, implies a need for a comprehensive volume on the particular subject of marine ecotourism. We hope that the book will allow readers to engage not only with the key issues of the topic, but also with the situations experienced by practitioners and researchers in different parts of the world.

Finally, we would like to acknowledge the support and advice of the following people in producing this book: David Bruce, Vincent Nadin and Maria Casado-Diaz (University of the West of England, Bristol) and The European Association of Leisure and Tourism Education (ATLAS).

Brian Garrod, Julie C. Wilson
Bristol
May 2002

References

Burton, R. (1998) Ecotour operators' responses to growth. *Journal of Sustainable Tourism* 6 (2), 117–42.
Fennell, D.A. (1999) *Ecotourism: An Introduction*. London: Routledge.
Hohl, A.E. and Tisdell, C.A. (1995) Peripheral tourism: Development and management. *Annals of Tourism Research* 22 (3), 517–34.
Weaver, D. (ed.) (2001) *The Encyclopedia of Ecotourism*. Wallingford: CABI.

Editors and Contributors

Brian Garrod holds a Bachelors degree from Portsmouth Polytechnic, a Masters degree from the University of East Anglia, and a PhD from the University of Portsmouth. He is presently Senior Lecturer in Environmental Economics and Associate Head of the School of Economics, University of the West of England, Bristol, UK (Brian.Garrod@uwe.ac.uk). He lectures in environmental economics and tourism economics and researches on the interface between environmental economics and sustainable tourism. His work is published in a range of journals, including *Annals of Tourism Research, Tourism Management,* the *International Journal of Sustainable Development* and *Environmental Politics*. He has twice been retained as a consultant to the World Tourism Organization (WTO) and has also worked as a consultant to the Organisation for Economic Cooperation and Development (OECD). Recent projects include updating and revising 'Agenda 21 for the Travel & Tourism Industry: Towards Environmentally Sustainable Development'.

Julie C. Wilson is presently Research Associate in Tourism with the Centre for Environment and Planning, University of the West of England, Bristol (Julie.Wilson@uwe.ac.uk). She has a BSc degree in Environmental Science and her PhD focused on the relationship between tourist place images and spatial behaviour. Her teaching and research interests include tourism and imagery, tourist behaviour, backpacker travel and ecotourism and she was recently Project Manager of the EU Interreg IIc project, Marine Ecotourism for the Atlantic Area (META-). She is currently working on various research projects, with funding from the Royal Geographical Society / HSBC Holdings, the British Council / NWO Netherlands, the UK Royal Society (Dudley Stamp Memorial Trust) and the UK National Trust.

Dr Mohamed El Ayadi is Assistant Researcher in Environment with the Seda para el Estudio de Humadales Mediterráneos (SEHUMED), Universidad de Valencia, Spain (elayadi@uv.es).

Dr Simon D. Berrow is Project Manager for the Shannon Dolphin and Wildlife Foundation, Kilrush, Co. Clare, Ireland (SDWF@oceanfree.net).

Dr Erlet Cater is Senior Lecturer in Tourism and Development in the Geography Department, The University of Reading, UK (e.a.cater@reading.ac.uk).

Ilika Chakravarty is a Research Scholar pursuing a PhD in the Department of Humanities and Social Sciences, Indian Institute of Technology (IIT), Powai, Mumbai, India (ilika_c@yahoo.com).

Dr Santiago Flores is Associate Teacher in Telecomunications at the Universidad Politécnica de Valencia, Spain (sflores@dcom.upv.es).

Elizabeth A. Halpenny is a PhD Candidate at University of Waterloo and a consultant with Nature Tourism Solutions, in Waterloo, Ontario, Canada (ehalpenny@sympatico.ca).

Salvador Herrera is an environmental consultant in Queretaro, México (chavo@hotmail.com).

Zena Hoctor is an independent tourism consultant in Ireland (zena@esatclear.ie).

Oscar Iroldi is Assistant Teacher at the Rochas School, Centro Politécnico del Cono Sur, Uruguay (tadi@adinet.com.uy).

Matthew McDonald is a Researcher in the School of Leisure, Sport and Tourism, Faculty of Business, University of Technology, Sydney, Australia (Matthew.McDonald@uts.edu.au).

Maryland Morant is Assistant Researcher in Environment with the Seda para el Estudio de Humadales Mediterráneos (SEHUMED), Universidad de Valencia, Spain (mmorant@uv.es).

Ghazali Musa is a visiting lecturer at Department of Tourism, University of Otago, New Zealand (ghazalimusa@pd.jaring.my).

Mark Orams, PhD is a Senior Lecturer and Director of the Coastal-Marine Research Group at Massey University at Albany, New Zealand (M.B.Orams@massey.ac.nz).

Christos Petreas is Managing Director of STADIA Ltd – Tourism, Development and Management Consultants, Athens, Greece (stadia@hol.gr).

Colin D. Speedie is Project Manager for the WWF-UK/The Wildlife Trust Basking Shark Project, c/o The Seawatch Charter, Penryn, UK (Colin.Speedie@btinternet.com).

Lola Teruel is Assistant Teacher in Environment and Tourism at the Universidad Politécnica de Valencia, Spain (dteruel@upvnet.upv.es).

Claudia Townsend is a freelance consultant and associate of the Centre for Responsible Tourism, University of Greenwich, Kent, United Kingdom (claudia.townsend@btopenworld.com).

Dr María José Viñals is Associate Researcher and Teacher in Environment and Tourism at the Universidad Politécnica de Valencia, Spain (mvinals@cgf.upv.es).

Dr Stephen Wearing is a Senior Lecturer in the School of Leisure, Sport and Tourism, Faculty of Business, at the University of Technology Sydney, Australia (s.wearing@uts.edu.au).

Introduction

JULIE C. WILSON and BRIAN GARROD

> 'The happiness of the bee and the dolphin is to exist. For man it is to know that and to wonder at it'. *Jacques-Yves Cousteau*

Defining marine ecotourism is not exactly a simple process. A great many definitions of ecotourism have been offered by scholars and practitioners worldwide, and although there is an emerging consensus that ecotourism is intrinsically nature-based, managed to be sustainable and includes an element of education and interpretation, there is still considerable controversy about how to define it. In this respect, it is perhaps easier to determine what is definitely not marine ecotourism. The least we can say of marine ecotourism is that it is a subset both of marine nature-based tourism and of sustainable marine tourism.

It is not our intention here to duplicate the many comprehensive analyses of ecotourism definitions that exist (see, for example, Diamantis, 1999; Fennell, 1999; Page & Dowling, 2002; Weaver, 2001). Indeed, in the first chapter of this book, Brian Garrod tackles the issue of defining marine ecotourism in an innovative way by analysing the views of 'experts' using the Delphi technique. All that is required at this stage, therefore, is for some of the elements that are widely agreed to be central to the concept of marine ecotourism to be outlined, without discussing in detail the (often very subtle) differences between the great many competing definitions of ecotourism that exist.

Some core themes are, however, clearly recurrent in most definitions of ecotourism. This has been noted by, among others, Beeton (1998) and Weaver (2001), who both conclude that there should be three themes observable in any activity claiming to be ecotourism. All of these three themes are recurrent throughout this volume and will be debated in various contexts in the chapters that follow.

The first characteristic of ecotourism is that it should be nature-based, i.e. the activities involved should focus predominantly on the natural environment. This is the fundamental 'eco' element of the term 'ecotourism' and it can potentially relate to the fauna of a destination area, the flora or, in many practical situations, both. This much is agreed. However, there are a number of more contentious issues involved in the requirement for ecotourism to be fundamentally nature-based. One is whether cultural or heritage-based attractions associated with natural areas should be included within the overall remit of the term 'ecotourism'. While some of the earliest and more famous definitions of the term (such as that of Cellabos-Lascuraín, in 1987) made an explicit attempt to include such activities, later definitions have elected not do so and there are clearly mixed opinions among ecotourism experts on this issue (as identified in Chapter 1). Meanwhile there is also an ongoing debate as to whether ecotourism can take place in highly modified environments such as plantation forests, farmland and urban environments (Higham & Lück, 2002).

Second, ecotourism should have a learning orientation, achieved through the processes of education and interpretation. This needs to be oriented towards changing visitors' behaviour in at least three different contexts: while undertaking the experience itself; while residing in the ecotourism destination area; and in the course of their daily life once they have returned back home. Furthermore, it needs to reflect not only learning about the local ecosystem, target species, culture, heritage and so forth but also about the global context within which these are set. Ecotourism is fundamentally about achieving change in the way people conceive of, contextualise and behave in natural environments, so that more sustainable relationships between humans and their environments may emerge. Education and interpretation are important mechanisms through which such change can be achieved.

Finally, and perhaps most importantly, ecotourism should be fundamentally underpinned by sustainability, through the application of principles of sustainable development. According to Beeton (1998), this should involve managing the physical stresses on the environment, including energy and waste minimisation, and wider environmental impacts. It should also look to manage the number of the people involved and the way they behave, including ecotourists' accommodation preferences, destination purchase decisions, the way they conduct themselves in the field and their spatial behaviour within fragile environments. Sustainability is not just about protecting the natural environment: it also extends to sustaining local livelihoods and ways of life for local communities. Sustainability is, of course, an essential concern for all kinds of tourism, not just ecotourism. The difference with ecotourism is that it takes its fundamental orientation from the principles of sustainability.

What Activities Might Potentially Constitute Marine Ecotourism?

Having raised some of the themes that need to be represented in order that ecotourism be considered genuine, we still have to ask what kind of activities might constitute ecotourism in the marine context. The chapters in this book have adopted a broad definition of 'marine' to encompass the foreshore, offshore and coast zones. In any case, there will be no clear distinction between these geographical zones in practice, these having a very close functional relationship in the marine ecotourism context.

Marine ecotourism activities can include watching whales, dolphins, other marine mammals and fish, birdwatching, scuba diving, beach walking, rock pooling, snorkelling, walking on coastal footpaths and sightseeing trips by surface boat, submarine and aircraft. It might also be considered possible that sea angling be included in this list, although this has attracted some controversy in the past. Land-based marine ecotourism activities, meanwhile, can include visiting Sea Life Centres, viewing coastal seascapes and, possibly, shore angling. While marine ecotourism is often portrayed as being based on wildlife attractions (and therefore sometimes classified as a subset of wildlife tourism), it is clear that there is also a huge variety of non-wildlife resources available to attract and interest marine ecotourists. For example, there is undoubtedly significant potential worldwide for the development of ecotourism that includes the unique cultural and heritage characteristics of coastal regions and communities. However, this cultural basis of ecotourism has often been overlooked by those providing, developing, researching and writing about marine ecotourism in favour of the natural environment aspects.

Marine ecotourism often focuses on what is sometimes called 'megafauna'. These are larger species of animal that are easy to observe, that appear regularly, predictably and in reasonably large numbers in specific locations, and therefore, that lend themselves to the development of viable commercial operations. Their sheer size can add to the appeal (e.g. whales) but species that are perceived to be physically attractive or particularly responsive to humans (e.g. dolphins) are also a big draw, as are some of the more dangerous species (e.g. great white sharks). Larger species tend to have relatively small populations, and require large territories (of relatively undisturbed ecosystems) to support viable breeding populations. It just so happens that such species are often excellent indicator species of the sustainable management of ecosystems and of environmental quality.

The international growth in marine-related tourism and recreation has also led to many operations emerging that are using the 'label' of marine ecotourism, when in practice they are clearly not oriented either in principle or practice towards achieving a more sustainable relationship with their natural, socio-cultural and economic environments. This has in the past been the case with certain scubadiving operations, for example, which have the appearance of ecotourism (in that they are nature-based and label themselves as ecotourism) but not the substance. Some scuba-diving providers, for example, allow divers to touch and thereby harm sensitive coral reefs, while others permit divers to take away coral as a souvenir of their day.

Characteristics of Marine Ecotourism Products and Providers

In understanding the different types of provision that come under the marine ecotourism banner, a distinction can be made between providers based on the level of emphasis on a particular species or group of species. Specialist providers actively promote the watching of a target species and use that species as a major part or the entire focus of the experience they offer. Specialist providers may also focus on more than one species, depending on the local abundance and reliability of sightings for any individual species. Marine ecotourism providers may also incorporate species into their products on an incidental basis, where a species becomes an incidental part of the experience on offer, based on opportunistic sightings. These opportunistic sightings may occur where species are migratory or occur seasonally, where sightings are inconsistent or unpredictable, or where the main focus does not include a particular species. However, in the latter case, where sightings are known to occur, they become part of the overall experience for the ecotourists (META- Project, 2000). In practice, many marine ecotourism enterprises are based on multiple species of marine wildlife. This might be because there are several key species available in a relatively small area or, more commonly, because of low levels of predictability of sightings. As such, products based on a combination of different species, as well as other non-wildlife resources, can be more viable. Furthermore, marine ecotourism experiences based on several different species in the same habitat area can give the opportunity for providers to maintain a more holistic educational dimension to the product.

The Relationship with Sustainability

The understanding of the concept of sustainability here equates to the 'strong' sustainability, as defined by Turner (1993) and discussed in the tourism context by Garrod and Fyall (1998). The term 'ecological sustainability', as used in the Australian National Ecotourism Strategy's definition of ecotourism, for example, is equated with the 'strong' sustainability of Turner's four interpretations ('very weak' through to 'very strong') (Turner, 1993). 'Strong' sustainability acknowledges that 'the substitutability of natural capital is severely limited by such environmental characteristics as irreversibility, uncertainty and the existence of 'critical' components' (Ekins & Simon, 1998: 169). The maintenance of the ecological processes on which biodiversity depends is therefore a core principle of ecotourism and a core objective for any marine ecotourism-based project. Those who initiate and develop ecotourism must hence operate within the capacity of the environment to absorb the impacts, as healthy species and habitats generally have stronger resilience to any negative impacts of tourism.

There are a number of persuasive reasons why ecotourism in the marine context is rather more complex to consider compared with terrestrial ecotourism. But what is so special about marine ecotourism that necessitates extra careful consideration? In Chapters 2 and 3, Erlet Cater and Julie Wilson both take up this issue in more detail, but it is worth stating here, in broad terms, why the marine context of ecotourism is especially important.

For one, if genuine ecotourism depends on accurate research and monitoring, then marine wildlife is extraordinarily difficult to study compared with terrestrial wildlife and is typically costly of time and resources, as it is highly mobile. As Julie Wilson observes in Chapter 3, this high level of mobility of many marine species can mean that marine tourism providers that are not operating according to the principles of genuine ecotourism are profiting from (and probably damaging) the very same wildlife resource that genuine providers are trying to access and sustain. This is certainly less of a problem in the context of terrestrial ecotourism. The same applies with regard to the issue of regulation, particularly when oriented to nature conservation. The widespread, often remote and therefore diffuse nature of many marine wildlife tourism interactions means that regulations are difficult, if not impossible, to police.

The Market for Marine Ecotourism

There is an emerging body of research on the characteristics of the principal markets for ecotourism. Perhaps the reason why not more has been done to date is that, especially on the demand side, there are various problems in identifying what counts as an ecotourism experience. In short, the typical 'ecotourist' simply does not exist. The reality is that ecotourism experiences are purchased by a very wide range of different types of tourist in combination with a variety of other kinds of tourist experience and across a wide spectrum of tourism settings. This extends from the small niche market of the 'dedicated wildlife watcher' at one extreme, to the larger but relatively under-researched potential market of the 'incidental' nature-based tourist at the other.

Given the difficulties of defining ecotourism, particularly as seen from the visitors' perspective, many studies of the market are limited to research on nature-based tourism, which is more easily identified from the supply-side perspective. In other words, many studies of the ecotourism market have simply defined ecotourists as people that visit destinations where the tourism product is predominantly nature based (e.g. countries like Costa Rica or Belize), national parks in general or designated protected areas. Other research has identified ecotourists as those who have specifically chosen nature-based tourism products.

On the whole, however, it has been argued that nature-based tourists have a particular set of demographic and psychographic characteristics (Blamey, 1995; Eagles, 1992; Fennell, 1999; Fennell & Smale, 1992; Wearing & Neil, 1999; Wight, 1996). In general, they tend to be

- older (particularly in the 35 to 54 age range),
- more highly-educated,
- slightly more likely to be male,
- prepared to pay rather more for their holiday,
- more frequent travellers,
- travelling in couples or small groups,
- drawn to tours offering a personalised service,
- motivated by intrinsic rather than extrinsic factors (i.e. principally interested in engaging with nature for its own sake),
- demanding of information and instruction on the destinations they visit,
- more likely to use 'adventure-type' accommodation such as cabins, lodges, inns, campsites, bed and breakfast, ranches,
- seeking to experience local conditions and sample customs, food and drink.

Pearce and Wilson (1995), however, identified a younger segment of wildlife watchers: younger than the average age for all visitors to New Zealand, likely to be fully independent travellers and likely to travel to more destinations and participate in a greater variety of activities. The wildlife watchers they studied were likely to spend more money in total, but less per day as their trips were generally longer, and they were more likely to backpack or camp. It was also argued that these tourists were seeking experiences in more remote locations.

Research by Orams (1999) on whalewatchers showed that they also tend to be higher-income and older-age groups and are relatively well educated (Forestall & Kaufman, 1996; Neil *et al.*, 1996). Interestingly, however, there are a greater proportion of females who participate in whalewatching (Neil *et al.*, 1996) and this was also observed by Reingold (1993) for ecotourists on the whole. This tendency has also been noted with dolphin swim participants (Amante-Helweg, 1995). Considering other marine tourism niche markets, scuba-diving tends to be dominated by males and younger-age groups (Davis & Tisdell, 1995; Orams, 1999), as are activities such as sailing, windsurfing and surfing, while those activities that are perceived to be more adventurous and higher risk are patronised mainly by males (Ewert, 1989).

Certainly, genuine ecotourism is still a niche market, although many people – particularly those in the tourist industry – tend to use the term in a much broader sense, referring to all nature-based tourism opportunities that, again, do not necessarily meet the principles of ecotourism in practice.

Why Might Marine Ecotourism be an Attractive Option for Peripheral Coastal Areas?

Marine ecotourism is most usually associated with out-of-the-way places – places that would be considered to be on the 'periphery'. This notion of peripherality might be viewed in geographic terms: that is, in terms of the physical remoteness of such locations. Alternatively, it might be viewed in economic terms: that is, in terms of their less inter-connected (and often dependent) relationship with the 'core' area of the country or region in which they are situated. This is not to suggest that marine ecotourism cannot take place, and be successful, in the 'core' areas. Indeed, marine ecotourism activities flourish in a number of major urban locations around the world, watching cetaceans from the west coast cities of the USA, Auckland in New Zealand and Vancouver, Canada, being well-known examples. At the same time, it is clear that one of the major features that attracts people to ecotourism in general is the feeling of remoteness that engaging in marine ecotourism activities can bring, of being 'at one with nature' in a wilderness setting, of leaving everyday life and taking 'time out'. These psychological draws would seem to apply equally well to marine ecotourism. Indeed, some of the most prestigious marine ecotourism locations lie in the remotest locations in the world, whalewatching in Antarctica perhaps representing the epitome of this general tendency.

Marine ecotourism thus has a particular affinity with remote and peripheral areas. Given the massive growth of interest in recent years in the activities covered by the term, there might therefore be good reason for local communities to look to marine ecotourism as a vehicle for their development. There are, indeed, a number of reasons why marine ecotourism might be considered a particularly attractive development option for peripheral coastal areas.

First, marine ecotourism, provided that it is conducted in ways that are fully consistent with the principles of sustainability, can offer peripheral areas a prospect for sustainable development. By developing marine ecotourism in a sustainable manner, economic development can be promoted in a way that is less likely to have detrimental impacts on the natural and cultural environments in which it takes place. As such, marine ecotourism may represent a form of development that addresses the needs of present and future generations at the same time. Of course, marine ecotourism that does not properly address the principles of sustainability, or only pays lip service to them, is unlikely to meet the needs of the local community in which it is set, be those needs expressed in economic, social or environmental terms. Marine ecotourism that is fundamentally damaging to the natural environment is unlikely to bring about sustainable development, even according to its weakest possible interpretation. Equally, marine ecotourism that is designed only with the intention of protecting the marine environment, or specific parts of it, will also fail to be truly sustainable unless it generates additional lasting benefits for the local community.

Second, a feature of many peripheral areas of the world, almost by definition, is that they contain many pristine coastal and marine resources – the kind that will be able to serve as a significant draw to marine ecotourists. In the core areas, meanwhile, potential marine ecotourism resources of this kind still tend to be few and far between, primarily because the past economic growth (which established the core areas as such) was invariably at the expense of the environment. By virtue of their location in the periphery, many coastal communities also have a distinctive maritime cultural heritage that might be used as the basis for marine ecotourism. These features give a comparative advantage to peripheral areas and arguably it is these distinctive features that such areas should seek to develop in attempting to catch up economically with the core areas.

A third reason why marine ecotourism may be seen as an attractive option for peripheral communities relates to the redeployment of resources and the need to invest in infrastructure. Many of the human-made resources required in order to develop marine ecotourism will either already be in place, for example harbour facilities and simpler forms of tourist accommodation, or could easily be converted to use in marine ecotourism activities. An example of the latter might be the conversion of traditional fishing vessels into whale- or dolphin-watching boats. Indeed, in many peripheral areas such resources are being freed up as a result of long-term economic decline in the location's traditional economic activities, such as commercial sea fishing and traditional seaside tourism. By redeploying some of these resources, the development of marine ecotourism can help soften the blow.

Marine ecotourism also has significant potential to employ resources that would otherwise remain unexploited and would therefore not contribute to local regeneration efforts.

Fourth, seasonality can be a particular problem in peripheral areas where tourism is conducted. Often there is a distinct peak of activity in the 'tourist season', when resources are at full stretch, followed by a similarly distinct trough in the 'off season', when resources are either under-employed or unemployed. This applies particularly to the tourism workforce, which is notoriously seasonal in its pattern of employment (Williams & Shaw, 1988). Many kinds of marine ecotourism activity, however, exhibit a more even seasonal pattern, some even having their peaks during the 'shoulder periods' of traditional tourism (usually spring and autumn). Bird-watchers, for example, will often be active in these seasons because various species of interest can be seen undertaking their annual migration. Marine ecotourism therefore represents a means of 'filling out' the tourism season, enabling a more efficient use of resources over the course of the year.

A fifth reason why marine ecotourism might be a particularly attractive option for peripheral coastal areas is that it represents a strategy of product diversification rather than market diversification. Some tourism destinations have sought to catalyse economic regeneration by attempting what is essentially a 'market diversification' strategy, which involves trying to find new markets for the existing tourism products offered by the destination, for example to try to attract 'winter sun' visitors to essentially 'summer holiday' destinations (Baum & Hagen, 1999). In many cases, however, this strategy has backfired, often because destinations have failed to recognise that effective market diversification also requires effective product diversification. This involves identifying different products, with different formulation, presentation, packaging and pricing to be attractive to 'out of season' tourists. Marine ecotourism, however, represents a form of product diversification. Moreover, many product diversification strategies that have been attempted in the past have suffered from the failing that they require significant levels of investment from essentially poor peripheral communities. Attempts to develop golf tourism in some destinations is a good example. Marine ecotourism, on the other hand, can be developed on a small, even micro scale, as indeed it is in many parts of the world, requiring only modest levels of capital investment.

A sixth possible reason why marine ecotourism might be seen as a particularly attractive development option for peripheral communities in coastal areas is that such destinations tend to suffer from the problem of 'expenditure leakage'. This is the tendency for expenditure on tourism to flow away from the destination in the form of payments for products, for example food and drink for tourists to consume in their accommodation or fuel for whalewatching vessels, which must be purchased from outside the immediate destination area. This means that a smaller proportion of tourism expenditure actually remains in the tourism destination to generate further 'multiplier effects' in the local economy. Small island economies, such as the Canaries and the Azores, are particularly prone to having high leakage factors because they are unable to produce so many of the things that are needed to support tourism activities themselves and are therefore forced to import them (Fennell, 1999). Studies suggest, however, that the consumption patterns of ecotourists tend to be much more like those of local residents, relying more heavily on locally-produced food and drink, locally-produced souvenirs and types of accommodation that can more easily be serviced from the local economy (Boo, 1991). They also tend to be 'higher spending' tourists, injecting more money into the local economy to be multiplied up than other forms of tourism (Wearing & Neil, 1999). Marine ecotourism can hence be associated with lower leakage factors and higher multiplier effects in the local economy than other forms of tourism.

Issues and Experiences of Marine Ecotourism

It is clear then that there are many threats to the likelihood of achieving genuine ecotourism in practice. In some cases, the threats are so acute that unless practices are modified straight away, irreversible damage may be done to the marine environment on which marine ecotourism depends for its success. For example, over-exploitation of marine resources through whaling has caused the collapse of many populations, sending some species to the brink of extinction. Marine ecotourism risks going the same way, with respect to some of the marine species it targets. There is evidence that populations of many species that are the focus of ecotourism enterprises are said to be declining, e.g. in the Moray Firth, Scotland, dolphin populations are said to be declining at 6% per annum, due to pollution, habitat disruption and tourism disturbance (Kelly, 2000).

In responding to the rapid growth in demand for marine ecotourism experiences, providers and managers are faced with immense risks. The high willingness to pay for 'swimming with cetaceans' experiences and an increasing international interest are already known about but the subsequent impacts of these current levels of interaction with marine wildlife species remain unknown. This is a serious concern, as the ecology of many of these species is often not fully understood. Better knowledge of the biology of the marine wildlife is essential, as often tourists interact with them at critical phases of their life cycle (e.g. mating or giving birth, feeding and migrating). Disruption of their behaviour during these critical periods, through the impacts of tourism, may prove to be an additional stress that wildlife populations cannot withstand.

Properly planned and managed though, marine ecotourism has the potential to exist while supporting marine wildlife conservation and research into the negative impacts that may arise. It should be thought of as both an ideal and a process by which that ideal can be realised. Certainly, it is likely that marine ecotourism will only be successful in the long term where management processes have been set in place that can ensure the conservation of the marine wildlife and natural resources on which the activity depends.

References

Amante-Helweg, V. (1995) Cultural perspectives of dolphins by ecotourists participating in a swim with dolphins programme in the Bay of Islands. MA Thesis, New Zealand: University of Auckland.

Baum, T. and Hagen, L. (1999) Responses to seasonality: The experiences of peripheral destnations. *International Journal of Tourism Research* 5 (1), 299–312.

Beeton, S. (1998) *Ecotourism: A Practical Guide for Local Communites*. Collingwood, Australia: Landlinks Press.

Blamey, R.K. (1995) The nature of ecotourism. Occasional Paper 21, Canberra, Australia: Bureau of Tourism Research.

Boo, E. (1991) Planning for Ecotourism. *Parks* 2 (3), 3–8.

Cellabos-Lascuraín, H. (1987) The future of ecotourism. *Mexico Journal* (Jan.), 13–14.

Diamantis, D. (1999), The concept of ecotourism: Evolution and trends. *Current Issues in Tourism* 2 (2&3), 93–122.

Davis, D. and Tisdell, C. (1995) Recreational scuba-diving and carrying capacity in marine protected areas. *Ocean and Coastal Management* 26 (1), 19–40.

Eagles, P.F.G. (1992) The travel motivations of Canadian ecotourists. *Journal of Travel Research* 31 (2), 3–7.

Ekins, P. and Simon, S. (1998) Determining the sustainability gap: National accounting for environmental sustainability. In P. Vaze (ed.) *UK Environmental Accounts: Theory, Data and Application* (pp. 147–67). London: HMSO.

Ewert, A. (1989) *Outdoor Adventure Pursuits: Foundations, Models, and Theories*. Scottsdale, AZ: Publishing Horizons.

Fennell, D.A. (1999) *Ecotourism: An Introduction*. London: Routledge.

Fennell, D.A. and Smale, B.J.A. (1992) Ecotourism and natural resource management. *Tourism Recreation Research* 17 (1), 21–32.

Forestall, P. and Kaufman, G. (1996) Whalewatching in Hawaii as a model for development of the industry worldwide. In K. Colgan, S. Prasser and A. Jeffery (eds) *Encounters with Whales '95* (pp. 53–65). Canberra: Australian Nature Conservation Agency.

Garrod, B. and Fyall, A. (1998) Beyond the rhetoric of sustainable tourism? *Tourism Management* 19 (3), 199–212.

Higham, J. and Lück, M. (2002) Urban ecotourism: A contradiction in terms? *Journal of Ecotourism* 1 (1), 36–51.

Kelly, A. (2000) North Sea dolphin colony 'may be wiped out within 50 years'. *The Independent Newspaper*, 10 April. Online document: http://www.independent.co.uk/story.jsp?story = 1729 (14 November 2001).

META- Project (2000) Marine Ecotourism for the Atlantic Area (META-): Baseline Report. Unpublished Research Report, University of the West of England, Bristol, UK.

Neil, D.T., Orams, M.B. and Baglioni, A.T. (1996) Effect of previous whalewatching experience on participants knowledge of, and response to, whales and whalewatching. In K. Colgan, S. Prasser and A. Jeffery (eds) *Encounters with Whales '95* (pp. 202–206). Canberra: Australian Nature Conservation Agency.

Orams, M.B. (1999) *Marine Tourism Development: Impacts and Management*. London and New York: Routledge.

Page, S. and Dowling, R. (2002) *Ecotourism*. Harlow: Pearson Education.

Pearce, D.G. and Wilson, P.M. (1995) Wildlife-viewing tourists in New Zealand. *Journal of Travel Research* 34 (2), 19–26.

Reingold, L. (1993) Identifying the elusive cultural tourist. *Tour and Travel News* 25 (Oct.), 36–7.

Turner, R.K. (ed.) (1993) *Sustainable Environmental Economics and Management: Principles and Practice*. London: Belhaven.

Wearing, S. and Neil, J. (1999) *Ecotourism: Impacts, Potentials and Possibilities*. Oxford: Butterworth-Heinemann.

Weaver, D. (ed.) (2001) *Ecotourism*. Milton, Australia: John Wiley & Sons.

Wight, P.A. (1996) North American ecotourists: Market profile and trip characteristics. *Journal of Travel Research* 34 (4), 2–10.

Williams, A.M. and Shaw, G. (1998) *Tourism & Economic Development: Western European Experiences*. London: Belhaven.

Section 1
Issues in Marine Ecotourism

Section 1
Issues in Marine Ecotourism

The first section of this book concentrates on outlining the proposed objectives of marine ecotourism, on considering the main principles adopted by proponents of ecotourism in addressing those objectives and on looking at some of the practical considerations that are consequently faced by those involved in planning, managing and regulating marine ecotourism. These are the key issues for marine ecotourism, which effectively establish its practical context and provide a setting within which specific experiences of marine ecotourism – which form the subject of the second section of this book – may be examined and analysed.

The first chapter, by Brian Garrod, begins by exploring different perceptions of the meaning of marine ecotourism, which is a term that evidently means different things to different people. Using an innovative methodology, comprising a combination of the Delphi technique and computer-aided content analysis, the chapter explores the various possible components of a definition of marine ecotourism and attempts to gain an understanding of which of these are central to the concept and which are not. Components such as management, sustainability and conservation are shown to be very strongly associated with the concept of marine ecotourism, while questions are raised regarding the desirability of including cultural manifestations within the remit of marine ecotourism and the need for marine ecotourism to take place in relatively undisturbed natural habitats. The chapter also presents some interesting insights into the merits and dangers of establishing a common definition of marine ecotourism.

Chapter 2, by Erlet Cater, then opens out the discussion by considering the range of negative, or 'backwash', effects of marine ecotourism, resulting from its positive, or 'spread', effects. Examples of marine ecotourism from around the world are used to illustrate the complex and integrated nature of these effects. The overall conclusion is that much greater collaborative efforts are required among those involved in all of the practical dimensions of marine ecotourism in order to identify areas of discord and concord among these effects. This will enable the various effects of marine ecotourism to be untangled, so that the negative impacts can more effectively be addressed and the positive impacts identified and built upon. Indeed, if marine ecotourism is to live up to its self-proclaimed reputation for sustainability, this must surely be the *sine qua non* for the success of marine ecotourism in meeting the ambitious objectives that have been set for it.

In Chapter 3, Julie Wilson considers the challenges for planning policy in enabling, promoting and regulating marine ecotourism. The nature of the marine environment is highlighted as a particular issue, its dynamic and sensitive nature representing perhaps the most problematic issue facing planners of marine ecotourism. For a number of reasons, the nature of the marine environment frustrates attempts to regulate and manage marine ecotourism activities. Marine ecotourism also faces an array of use-conflict and sectoral issues that raise particular dilemmas for planning and management. These are important issues that set marine ecotourism apart as a special area of concern and highlight the need to study ecotourism specifically in the marine context. Such issues simply do not arise with equivalent ecotourism activities in purely terrestrial settings.

The theme of regulation is subsequently taken up in Chapter 4, by Simon D. Berrow, which undertakes an assessment of the framework, legislation and monitoring required for developing genuinely sustainable whalewatching. Watching whales and dolphins is an economically important activity in many countries of the world, yet the extent to which it can have a detrimental effect on the behaviour and habitat of the species involved has not been fully assessed. Meanwhile a wide variety of regulatory frameworks have been used to manage whalewatching around the world, ranging from purely voluntary codes of conduct, to formal legislation, most often involving a blend of these two extremes. The chapter concludes that the best opportunities for effective regulation exist where whalewatching activities can be controlled through formal licensing systems. Marine Protected Areas (MPAs) are most relevant where critical habitats and resources have been identified. In the meantime, models of best practice, including codes of conduct and accreditation schemes, need to be developed for application within these overall frameworks of regulation.

Chapter 5, by María José Viñals and colleagues, discusses the use of recreational carrying capacity as a tool for managing tourism in wetland environments. Wetland areas have not traditionally been considered in terms of their role as places for leisure and tourism, yet leisure- and tourism-related activities in such areas are becoming increasingly popular. The establishment of recreational carrying capacity is therefore seen as an important first step in developing a systematic approach to managing tourism in wetland areas. This task is, however, not without its difficulties and this chapter attempts to address some of these by suggesting a step-by-step methodological approach.

Chapter 1

Defining Marine Ecotourism: A Delphi Study

BRIAN GARROD

Introduction

The United Nations has designated 2002 the International Year of Ecotourism (IYE 2002). A year of intense discussion, both about the concept and the practice of ecotourism, is widely anticipated. Yet even with the first warning shots of the contest having been fired towards the end of 2001, it has become increasingly clear that a fundamental problem exists: participants simply do not agree about what they believe ecotourism to be, either in principle or in practice. The lack of a widely agreed or accepted definition of ecotourism must raise considerable doubts as to whether those involved in the IYE 2002 will be able to come to a workable common view on what ecotourism should be attempting to achieve and how good practice might best be promoted. With so great a potential for talking at cross-purposes, it is likely that the proceedings of the IYE 2002 will shed much heat but little light.

That those involved in ecotourism at various levels of government should fail to have a common view on how it should be defined would come as no great surprise to academics and practitioners of tourism, who have long recognised the problem. Particular areas of contention include whether ecotourism should be based conceptually on relatively 'weak' or relatively 'strong' versions of sustainability, whether ecotourism should be viewed as a subset of 'nature-based tourism', 'sustainable tourism' or even 'wildlife tourism' and whether its application should be restricted to activities that are deemed 'non-consumptive'. Useful reviews of such issues are to be found in Orams (1995), Blamey (1995, 1997), Wall (1997), Burton (1998), Tremblay (2001) and, most recently in Page and Dowling (2002). Diamantis (1999), meanwhile, presents and analyses some 15 proposed definitions of the term 'ecotourism', dating from between 1987 and 1997.

One important reason why academics and practitioners have been keen to develop agreed definitions of ecotourism is to establish a means by which genuine ecotourism can be distinguished from activities that merely employ the term (perhaps cynically) as a marketing buzzword. The existence and growth of such counterfeit ecotourism holds two main dangers. First, the irresponsible and unsustainable practices involved may, if they become widely known about, tarnish the hard-fought reputation of the ecotourism 'brand' as representing genuinely sustainable tourism. Second, such activities may damage the natural resources that genuinely sustainable ecotourism depends upon and is trying to maintain.

This chapter outlines the findings of study of experts from a range of backgrounds and located across the EU's 'Atlantic Area', based on a survey technique known as the 'Delphi method'. The focus of the study was on how ecotourism might best be defined in the marine context, if indeed closer definition of the concept is considered to be helpful in terms of achieving its practical objectives. By initiating and coordinating the deliberations of a panel of experts from across the EU Atlantic Area, and by drawing lessons from this work, the study aimed to establish the basis for a transnational view regarding what marine ecotourism is considered to be in principle. This, it was hoped, would lead to a better understanding of what needs to be done in practice to achieve genuinely sustainable marine ecotourism in the EU Atlantic Area context.

The Delphi Technique

The Delphi technique has been defined as follows:

> A systematic method of collecting opinions from a group of experts through a series of questionnaires, in which feedback on the group's opinion distribution is provided between question rounds while preserving the anonymity of the respondent's responses. (Helmer, 1972; cited in Masser & Foley, 1987)

The basic rationale of the Delphi technique is to elicit expert judgement on issues or problems that are highly complex and essentially subjective in nature, requiring the use of a substantial degree of expertise on the part of those addressing them. Such issues or problems cannot easily be dealt with using conventional questionnaire- or interview-based survey techniques. Indeed, past experience has shown that simply asking experts for their opinions about complex issues or problems tends to yield unreliable results. One reason is that complex questions invariably have complex answers, yet experts do not necessarily have (or take) the time to cogitate at length on the issues raised in the survey, to think deeply about the problem under consideration or to develop their answers thoroughly. Nor do they necessarily test their ideas out by exposing them to rigorous peer evaluation. This can lead to unreliable judgements being made about the issues covered in the survey. In short, conventional survey techniques have a tendency to collect 'snap judgements' on the complex issues the researcher is trying to study, rather than the carefully considered, in-depth, peer-evaluated expert opinions that are required if such complex problems are to be meaningfully addressed.

One possible means of addressing the shortcomings of conventional surveys when dealing with expert subjects might be to hold a seminar or symposium, to which a range of experts would be invited and encouraged to address the issues at hand in a more interactive arena. Relevant questions could be sent to the participants in advance, so that on the day of the event the experts would be able to air their views, think them through in relation to those put forward by their peers and make a deeply-considered final response. This final response may even be solicited through a follow-up questionnaire administered shortly after the event has taken place, giving the participants further time for thought and reflection.

The major problem with this approach, however, is that in attempting to address the inherent problems of conventional survey instruments, potentially even more serious concerns about reliability are encountered. A particular concern would be the potential influences of personality, institutional allegiances and peer pressure on the experts' expressed views. Some participants may, for example, be unwilling to depart in public from the conventional wisdom of their discipline or profession. Others may be reluctant to adopt an opposing stance to the official view of the organisation that employs or sponsors them. Others again may feel unhappy about expressing views that might leave them isolated in a polarised public debate. There may even be a tendency for participants to court controversy where none really exists, simply because that is what they think the organisers expect and the nature of the event demands.

The Delphi technique attempts to avoid – or at least to minimise – these potential biases by allowing a small but carefully chosen panel of expert participants to address the issues at hand in a structured, deep and anonymous way. Based on a description of the technique by Richey *et al.* (1985), Table 1.1 identifies the basic steps involved in a Delphi study. The principal features of the approach are as follows.

First, the technique uses a small panel of experts, selected purposively on the basis of their expertise in subjects related to the issues that are to be addressed through the Delphi study. The size of the panel is not normally considered to be a critical issue (Smith, 1995); what is considered more important is that the panel is suitably balanced in terms of the background, interests and expertise of its members. Second, the technique is iterative, with questioning of panel members taking place over a number of 'rounds'. This enables participants to suggest a tentative response to preliminary rounds and to refine their responses in the course of subsequent rounds. Third, the technique involves indirect, rather than direct, interaction between the panel members. Rather than meeting face to face, participants meet 'virtually' through the cyclical iteration of the responses of panel members. In this way, participants can learn how others have answered the questions at hand, understand why they have done so, build wider considerations into their own responses and adapt them accordingly. Fourth, the interaction process is anonymous, thereby minimising the potential for interpersonal or group biases to creep into participants' responses. Fifth, the technique takes place over an extended period of time, giving participants

Table 1.1 Basic steps in the Delphi technique

1.	Choose the members of the coordinating group
2.	Develop criteria for evaluating potential candidates for the expert panel
3.	Identify potential candidates
4.	Request candidates' participation (perhaps in person by a prestigious individual)
5.	Finalise panel composition and set criteria for the acceptable balance of the panel membership
6.	Identify issues to be considered and develop the initial ('scoping') questionnaire
7.	Send the first questionnaire
8.	Collate the responses and check for balance of the panel (if the panel becomes unacceptably unbalanced the study should not continue further)
9.	Develop the second (convergence) questionnaire, incorporating all new input. Use a numerical scale or ranking system to calibrate responses
10.	Send the second questionnaire
11.	Collate the responses and check for panel balance
12.	Undertake further rounds as deemed necessary (e.g. until an acceptable degree of consensus is achieved)
13.	Send a summary of final results to all respondents
14.	Apply the resulting judgement(s) to solve the problem(s) being addressed through the Delphi study

the opportunity to think deeply about their responses and to shape them as the study progresses through successive rounds.

The Delphi technique was developed by the RAND Corporation in the 1950s, finding its earliest application in the field of military strategy (Dalkey & Helmer, 1963). However, it was not until the 1960s and 1970s that the Delphi technique became more widely recognised, being applied particularly in an area of analysis known as 'technological forecasting' (see, for example, the forecasts of technological advances in medical science into the 1990s by Teeling-Smith [1971]).

In the field of tourism, the Delphi technique has occasionally been used in forecasting; most often to forecast demand (see, for example, Kaynak & Macaulay, 1984; Liu, 1988). Indeed, during the 1970s and 1980s this was the predominant use of the Delphi technique in the tourism context. In the 1990s, however, the Delphi technique began to be used by researchers in more qualitative ways. A study by Green *et al.* (1990), for example, attempted to employ the Delphi technique to identify the range and extent of potential environmental impacts associated with the development of a new tourism attraction. Similarly, a set of three studies undertaken by Pan *et al.* (1995) considered a range of qualitative issues relating to destination marketing. Garrod and Fyall (2000) applied the Delphi technique to investigate a number of qualitative issues relating to the sustainable management and funding of heritage tourism attractions in the UK. Miller (2001), meanwhile, employed the Delphi technique to generate and develop a series of indicators for sustainable tourism.

An important feature of these more qualitative applications of the Delphi technique is that they do not necessarily seek consensus on a particular issue or problem. Indeed, if experts' views widen as a result of the Delphi process, this may be a valid and helpful finding in itself. Today the Delphi technique is seen more as a means of creating a productive 'think tank' within which problems can be mulled over, developed and thought about creatively. A final group consensus is neither sought nor necessarily considered appropriate given the nature of the issues being considered. As Miller (2001: 356) remarks: 'it is perhaps symptomatic of tourism's multidisciplinary nature, that even with no ego involvement ... respondents may not feel able to achieve ... agreement'.

Another important feature of many of the more qualitative uses of the Delphi technique undertaken since the early 1990s is that the study is not generally seen as an end in itself; rather it is seen as a useful supplement to other, perhaps more statistically rigorous techniques. Hence a Delphi study might be used to follow up on the results of a quantitative survey with a subset of respondents, enabling more qualitative aspects of the subject matter to be developed.

The Delphi technique is undoubtedly controversial and has attracted a great deal of academic debate since it was first applied. Among the best known critiques of the Delphi technique are those by Hill and Fowles (1975) and Sackman (1975), while Linstone and Turoff (1975) set out some of the strengths of the technique. A more rounded review of the Delphi technique, meanwhile, can be found in Rotundi and Gustafson (1996).

Design of the Delphi Survey

The Delphi survey reported on in this chapter was coordinated by a small group of researchers based at the University of the West of England, Bristol, under the general auspices of a project entitled 'Marine Ecotourism for the Atlantic Area (META-)', funded under the EU Interreg IIc Programme. A panel was formed comprising 15 experts drawn from the three countries with partners in the project, namely Ireland, the UK and Spain (the Canary Islands) where parallel projects were being conducted. It was felt that this size of panel would be large enough for differing views to emerge among the panel members, while being small enough to allow the questionnaires to be turned around in a timely manner. Experts in the field of ecotourism tend to operate in an international context where English is widely used. It was therefore considered appropriate to use English as a common working language.

Establishing and maintaining a well-balanced panel is considered to be especially important to the success of any Delphi exercise (Wheeller *et al.*, 1990). Any Delphi panel should therefore be able to demonstrate a good balance of professional and academic backgrounds, personal and professional interests, and national locations. Candidate panel members were therefore asked to fill in a confidential personal profile. This provided details of their educational background, professional qualifications, career experience, marine ecotourism interests and their particular areas of expertise related to marine ecotourism.

The following criteria were then used to judge the suitability of the experts to sit on the panel:

- practical experience of developing marine ecotourism or a demonstrated interest in doing so;
- at least five years professional experience in a relevant field; and
- good knowledge either of a META- parallel project location or a particular type of marine ecotourism (cetacean watching, seabird watching, scuba diving, etc.).

Candidates who were considered not to meet one or more of these criteria would be thanked for their interest but not be invited to sit on the panel. This selection process helped to ensure that the experts were well versed in ecotourism issues and possessed the kinds of knowledge and skills that would be required in order to answer the questions set out in the Delphi questionnaires.

Meanwhile, three criteria were set for maintaining balance of the panel:

- the panel should comprise no more than 15 members;
- the panel should include no more than one-third of the members sharing the same profession, academic background or national location; and
- the panel should have as little duplication of members' fields of interest and expertise as possible.

Previous Delphi studies in the field of tourism have been criticised because of the tendency for the panel to become unbalanced as participants dropped out of the iterative survey process (Wheeller *et al.*, 1990). There are many possible reasons why panel members might wish not to continue with the project, some of which may be related to the conduct of the project itself (e.g. survey rounds not taking place in a timely manner so that panel members become bored with the process) or unconnected to the conduct of the project (e.g. other work commitments that must take precedence). It was therefore decided that if the expert panel should more than once fail to meet any one of the criteria for balance, no further rounds of the survey would be attempted and the project would effectively be terminated at that point. To the author's knowledge, this is the first time that such 'quality control' measures have been put into place before commencing on a study using the Delphi technique.

The First Round

The first round of the Delphi study presented the panel members with a wide range of possible working definitions of ecotourism (see Table 1.2). It was hoped that this would help focus their minds on the differences and similarities in fundamental thinking on what ecotourism should be considered to be in principle – and therefore what it should be aiming to achieve in practice. The definitions were drawn from the academic and professional literature to illustrate a range of

Table 1.2 Initial ten definitions of ecotourism

1.	'Any form of tourism which is based on the natural ecological attraction of a country, ranging from snorkelling off coral reefs to game viewing in savanna grasslands.' (Cater, 1992)
2.	'Nature-based tourism that involves education and interpretation of the natural environment and is managed to be ecologically sustainable.' (Commonwealth Department of Tourism, 1992)
3.	'Tourism that consists of travelling to relatively undisturbed or uncontaminated natural areas with the specific objective of studying, admiring and enjoying the scenery and its wild plants and animals, as well as any existing cultural manifestations.' (Ceballos-Lascuraín, 1987)
4.	'Purposeful travel to natural areas to understand the culture and natural history of the environment, taking care not to alter the integrity of the ecosystem, while producing opportunities that make the conservation of natural resources beneficial to local people.' (Ecotourism Society, 1994)
5.	'A sustainable form of natural resource based tourism that focuses primarily on experiencing and learning about nature, and which is ethically managed to be low impact, non-consumptive, and locally orientated (control, benefits and scale). It typically occurs in natural areas, and should contribute to the conservation and preservation of such areas.' (Fennell, 1999)
6.	'Ecotourism can contribute to both conservation and development and involves, as a minimum, positive synergistic relationships between tourism, biodiversity and local people, facilitated by appropriate management.' (Ross & Wall, 1999)
7.	'That kind of tourism which is: (a) based on relatively undisturbed natural areas, (b) non-damaging, non-degrading, (c) a direct contributor to the continued protection and management of the natural areas used, (d) subject to an adequate and appropriate management regime.' (Valentine, 1993)
8.	'Ecotourism consists of three core criteria: the primary attraction is nature-based (such as flora and fauna, geological features), with cultural features constituting a secondary component; the emphasis is on the study and/or appreciation of the resource in its own right; and the activities of the tourists and other participants are benign with respect to their impact upon the physical and cultural environment of the destination. Ecotourism … should be coherent with the notion of sustainable tourism by adhering to the carrying capacities of the destination and being acceptable to, and supportive of, host communities.' (Weaver, 1999)
9.	'The planned practice of tourism in which the enjoyment of nature and learning about living beings and their relationship with the environment are brought together; it is an activity which does not result in a deterioration of the environment and which promotes and supports the conservation of natural resources, thereby producing economic benefits which reach most social strata of the population in such a way that a sustainable horizontal development is achieved. Moreover, real ecotourism promotes justice for people and for nature.' (Evans-Pritchard & Salazar, 1992)
10.	'Where an individual travels to what he or she considers to be a relatively undisturbed natural area that is more than 40 km from home. The primary intention being to study, admire, or appreciate the scenery and its wild plants and animals, as well as any existing cultural manifestations (both past and present) found in these areas.' (Blamey, 1997)

Table 1.3 Initial choices of the panel

Definition	*No. choosing*
1	1
2	2
3	0
4	1
5	4
6	3
7	0
8	1
9	2
10	0
Total	14

possibilities, with the definitions included varying in length, style and content. Some were adapted slightly so that they made more sense as 'stand alone' definitions.

The first-round questionnaire had three main sections. In the first, panel members were asked to choose the definition with which they felt most comfortable in the context of marine ecotourism in the EU Atlantic Area. Fourteen responses were received following a reminder sent by e-mail two weeks after the deadline for responses. Respondents were also asked to provide a brief written justification for their choice. Table 1.3 presents the initial choices of the 14 respondents.

Second, panel members were asked how, if at all, they would improve upon the definition they had chosen. Space was provided on the questionnaire to allow respondents either to amend an existing definition or suggest their own. Panel members were then asked to explain briefly what changes they made, why and to state the sources of any additional infromation used.

Ten panel members chose to provide new or adapted definitions, the other four stating that they were content with their chosen definition from the original list of ten. These additional definitions were collected together and numbered 11 to 20, so that panel members could be presented with an expanded list of 20 definitions from which to choose in the second round (see Table 1.4).

Third, panel members were asked for their views on the value of definitions of ecotourism, in particular the reasons why defining ecotourism might be considered either helpful or unhelpful in achieving its objectives. A box was provided for respondents to provide written comments, which were to be collated and returned to the panel for further consideration in the second round. These comments are presented in Table 1.5:

Following standard procedure in Delphi studies, the panel member who did not respond to the first-round questionnaire was contacted and thanked for their help but not invited to take part in further rounds.

Table 1.4 Additional definitions provided by panel members

11.	'Any form of tourism that is based on the ecology of natural environments'.
12.	'Nature-based tourism that involves education and interpretation of the natural environment and is managed to be ecologically sustainable while producing opportunities that help to make the conservation of natural resources beneficial to local people.'
13.	'Nature-based tourism that involves education and interpretation of the natural environment and is managed to be ecologically sustainable. Secondary considerations, refining the principal definition may include: ethically managed, local control and economic benefit and respect for minority/indigenous cultures, but are not necessarily part of the definition. Eco-tourism should also reflect the potential for other ways of learning/respecting the natural environment and promoting tourism such as through music, poetry and art.'
14.	'*Responsible* travel to natural areas to understand *admire and enjoy* the culture and natural history of the environment taking care not to alter the integrity of the ecosystem, while producing opportunities that make the conservation of natural resources beneficial to local people.'
15.	'A form of natural resource based tourism that focuses primarily on experiencing or learning about nature, and is therefore ideally managed to be environmentally sustainable. It typically occurs in relatively undeveloped (natural) areas, and should contribute to the conservation and preservation of such areas while supporting the development of the region's economy.'
16.	'The development and use of a natural resource to enhance local culture and tourism in a long term sustainable way through education and together with prudent and sensible management of the asset for future generations.'
17.	'A sustainable form of natural resource based tourism that focuses primarily on experiencing and learning about nature, and which is ethically managed to be low impact, non-consumptive, and locally orientated (control, benefits and scale). It typically occurs in natural areas, and should contribute to the conservation, preservation, reclamation and restoration of such areas.'
18.	'Ecotourism can be defined as a low impact travelling, lodging and respectful moving in a foreign (known or unknown) territory. The tour operators are supposed to control that those tourists don't get off the beaten track and wisely guide and educate in order to make people more aware and sensitive towards mother nature.'
19.	'Ecotourism can contribute to both conservation and development and involves as a minimum positive synergistic relationships between tourism, biodiversity and local people facilitated by appropriate management and education.'
20.	'Nature-based tourism that involves education and interpretation of the natural environment and is managed to be ecologically sustainable.'

The Second Round

The second-round questionnaire was sent out as soon as the last first-round questionnaire was returned. The questionnaire was accompanied by a feedback document setting out the various responses to the first-round questionnaire. This was done in order to enable panel members to see how other panel members had responded to the first round questionnaire and to identify where their own

Table 1.5 Comments on the value of definitions of ecotourism

Reasons why definitions of ecotourism can be HELPFUL:	
1.	As a basis for further debates about the ethics and applications of the term.
2.	As a means of putting the idea of ecotourism across to local people.
3.	To guide in discussions that can assist in effective planning and policy-making.
4.	To help those working in ecotourism to put their work into context.
5.	As a means of communicating the objectives of ecotourism projects.
6.	As a means of focusing the input of disparate bodies, helping them to work toward a common aim.
7.	To assist in the initial audit of an ecotourism project that is going to be launched.
8.	As a reference point to ensure the relevance of efforts in making ecotourism happen.
9.	As a constant reminder for everyone of the reasons for creating ecotourism projects.
Reasons why definitions of ecotourism can be UNHELPFUL:	
1.	As the concept becomes more widely known and fashionable there is a danger that large organisations will use it merely as a 'green label' for their existing activities, cynically exploiting people's good will toward the idea.
2.	Obsession with finding the best definition of ecotourism can obscure the practical side of ecotourism.
3.	The appropriate definition of ecotourism may depend on circumstances, according to people's priorities.
4.	Too narrow a definition of ecotourism can be unhelpful if it does not allow the concept to be applied in certain situations or if certain stakeholder groups are excluded by it.

responses fitted within the overall spread of views. In this way, panel members could judge where (and, to some extent, why) their responses differed from others and to amend their second-round responses accordingly.

Eleven responses were received to the second-round questionnaire. However, while the criteria for balance relating to the education of panel members, national location and areas of panel member interest and expertise were all met, the criterion relating to the panel members' career background was not (there were proportionally too many ecotourism providers).

With regard to the definition of ecotourism in the context of marine ecotourism in the EU Atlantic Area, the second-round questionnaire asked the panel members to reconsider their preferred definition of ecotourism and to choose again from the expanded list of 20 definitions. Panel members were reminded that they were entitled to stick to their original choice but that they had to provide a brief justification for whatever choice they took. Table 1.6 presents the new distribution of preferred choices after the second Delphi round:

In respect of the question about the value of definitions of ecotourism, panel members were asked to select the two factors that they felt to be first and second

Table 1.6 Choice of definition in second round

Definition	*No. choosing*	*Definition*	*No. choosing*
1	0	11	1
2	0	12	1
3	0	13	0
4	1	14	1
5	0	15	1
6	1	16	0
7	0	17	1
8	0	18	1
9	2	19	1
10	0	20	0
		Total	11

most important from each list. They were also given a further opportunity to add to each list should they wish to do so. Table 1.7 presents the results of this exercise.

The third Delphi round achieved only ten responses and the criteria for balance of the panel continued not to be met. It was, therefore, considered appropriate not to continue with the study from this point, and to include only the data collected in the first two rounds in any further analysis.

Content Analysis of the Definition Sets

In order to learn more about the characteristics of those definitions favoured by the Delphi panel, and to understand why they were preferred to others that were not, a content analysis of the four definition sets collated in the course of the two Delphi rounds was conducted. QSR N-Vivo, a text-handling software package, was used to assist in this process. The four definition sets were as follows:

Set 1: the initial set of ten definitions chosen by the research team ($n = 10$);
Set 2: the panel's preferred choice from these ten definitions in the first round ($n = 14$);
Set 3: the panel's preferred choice from the expanded list of 20 definitions in the first round ($n = 14$); and
Set 4: the revised set, accounting for any changes the panel wished to make in the second round ($n = 11$).

Table 1.7 Reasons why definitions of ecotourism can be helpful and unhelpful

No.	*Explanation*	*First choice*	*Second choice*
Reasons why definitions of ecotourism can be helpful			
1.	As a basis for further debates about the ethics and applications of the term.	0	0
2.	As a means of putting the idea of ecotourism across to local people.	4	1
3.	To guide in discussions that can assist in effective planning and policy-making.	0	2
4.	To help those working in ecotourism to put their work into context.	1	0
5.	As a means of communicating the objectives of ecotourism projects.	0	0
6.	As a means of focusing the input of disparate bodies, helping them to work toward a common aim.	1	3
7.	To assist in the initial audit of an ecotourism project that is going to be launched.	0	1
8.	As a reference point to ensure the relevance of efforts in making ecotourism happen.	1	2
9.	As a constant reminder for everyone of the reasons for creating ecotourism projects.	2	2
10.	To have an exact and accepted definition.	1	0
11.	To promote justice for people and nature.	1	0
	Total:	11	11
Reasons why definitions of ecotourism can be unhelpful			
1.	As the concept becomes more widely known and fashionable there is a danger that large organisations will use it merely as a 'green label' for their existing activities, cynically exploiting people's goodwill toward the idea.	2	3
2.	Obsession with finding the best definition of ecotourism can obscure the practical side of ecotourism.	3	0
3.	The appropriate definition of ecotourism may depend on circumstances, according to people's priorities.	0	3
4.	Too narrow a definition of ecotourism can be unhelpful if it does not allow the concept to be applied in certain situations or if certain stakeholder groups are excluded by it.	6	4
5.	If they don't promote justice for nature and people.	0	1
	Total:	11	11

For the purposes of the determining the characteristics of each definition set, a definition that was chosen by two panel members was included twice in the counting process, a definition chosen by three panel members was included three times, and so on.

The following characteristics were used to classify the content of the definitions:

- whether or not the definition recognised a need for interpretation and/or education as a component of the ecotourism experience (node = 'education');
- whether or not the definition recognised a need to manage ecotourism (node = 'management'), suggesting that ecotourism, if poorly managed, might actually harm the things that the ecotourists are paying to see;
- whether or not the definition argued that ecotourism must be sustainable (node = 'sustainable'), implying that ecotourism is to be viewed as a subset of 'sustainable tourism';
- whether or not the definition identified the natural areas and species that are the subject of marine ecotourism as tourist 'attractions' (node = 'attraction'), suggesting an instrumental use of nature as a tourism 'resource';
- whether the definition recognised the need for marine ecotourism (simply) to benefit locals (node = 'benefits to locals') or whether the definition went further to recognise the need for local stakeholders to have an active participation in the planning and management of marine ecotourism (node = 'involving locals');
- whether or not the definition focused on ecotourism primarily as a tool of conservation (node = 'conservation') as opposed to, for example, being primarily an instrument of economic development;
- whether the definition recognised a cultural component to ecotourism (node = 'cultural), as opposed to viewing ecotourism as being essentially nature-based (node = 'nature-based');
- whether or not the definition considered ecotourism only to be possible in (relatively) undisturbed areas (node = 'undisturbed'), as opposed to heavily human-influenced environments such as forested or urban areas;
- the word length of the definition (nodes = 'size 0–30', 'size 31–60', and 'size 61+'), reflecting possible preferences for 'short and snappy' definitions or longer, more detailed definitions; and
- the style in which the definition is written, ranging from the generally descriptive (node = 'style 1') to the heavily criteria-based (node = 'style 5').

All of these nodes except for those relating to style could be analysed objectively by examining the definition – key terms were either mentioned or they were not; length was assessed by means of word counts. The set of nodes relating to style was analysed by two researchers attempting the classification independently, then coming together to discuss areas of similarity and difference, ultimately to decide on a mutually agreed distribution of definitions among the five possible nodes. Table 1.8 presents the results of this content analysis.

Table 1.8 Analysis of the Content of Definition Sets

Node	*Set 1*	*Set 2*	*Set 3*	*Set 4*	*Set 1 (%)*	*Set 2 (%)*	*Set 3 (%)*	*Set 4 (%)*
Education	7	11	12	9	70.0	78.6	85.7	81.8
Management	6	12	12	8	60.0	85.7	85.7	72.7
Sustainable	7	13	11	9	70.0	92.9	78.6	81.8
Attraction	3	3	4	3	30.0	21.4	28.6	27.3
Benefits to locals	5	11	10	9	50.0	78.6	71.4	81.8
Involving locals	2	7	5	3	20.0	50.0	35.7	27.3
Conservation	6	11	8	9	60.0	78.6	57.1	81.8
Cultural	4	3	4	2	40.0	21.4	28.6	18.2
Nature-based	5	8	9	4	50.0	57.1	64.3	36.4
Undisturbed	3	1	1	1	30.0	7.1	7.1	9.1
Size 0–30	3	6	3	3	30.0	42.9	21.4	27.3
Size 31–60	5	6	8	6	50.0	42.9	57.1	54.5
Size 61+	2	3	3	2	20.0	21.4	21.4	18.2
Style 1	1	1	1	1	10.0	7.1	7.1	9.1
Style 2	2	1	2	1	20.0	7.1	14.3	9.1
Style 3	2	5	5	4	20.0	35.7	35.7	36.4
Style 4	2	1	1	2	20.0	7.1	7.1	18.2
Style 5	3	7	5	3	30.0	50.0	35.7	27.3
n =	10	14	14	11	100.0	100.0	100.0	100.0

The first three criteria – whether ecotourism should include interpretation and education of the environments in which it takes place, whether it should be subject to a management regime and whether it should aim to be sustainable – all received strong support from the panel members. These criteria are probably the most widely accepted in the literature (see Garrod, 2002), so that the panel should strongly support these criteria should really come as no surprise. However, it is significant that in all three cases the panel had by the end of the second round indicated an even stronger preference for those criteria to be included in their chosen definition.

Meanwhile, the panel was rather non-committal in respect of whether the definition portrayed the natural areas and species involved in ecotourism as 'attractions', for instrumental use by the tourism industry or as having intrinsic value in their own right. Three of the original ten definitions included this aspect, the figure falling slightly to just over 27% by the end of the second round. Panel members were evidently neither strongly attracted to this aspect of possible definitions of ecotourism, nor particularly keen to avoid it.

With regard to the issue of whether locals should benefit directly from ecotourism, more clarity is apparent from the content analysis. While 50% of the definitions in the original set acknowledged the need for locals to share in the economic benefits of ecotourism, by the end of the second round this figure had risen to over 80%. The argument would seem to be that local people must receive appropriate economic benefits from ecotourism if it is to win their commitment, both as a principle and operationally (for example, in terms of encouraging them to abide by whatever management structures are put in place). There is indeed a great deal of evidence to support this view. Drake (1991), for example, recounts the case of the Amboseli National Park in Kenya, where local Masai people became tired of not receiving the economic benefits promised to them in return for abiding by strict regulations on hunting and the grazing of animals. The result was defiant and purposeful encroachment on the reserve in order to hunt its wildlife and to allow their cattle to drink at the water source. Meanwhile Inskeep (1999) argues that making local people direct beneficiaries of ecotourism has had an important enabling effect on the development of ecotourism in many parts of the world, including the Communal Areas Management Programme for Indigenous Resources (CAMPFIRE) in Zimbabwe and the Administrative Management Design for Game Management (ADMADE) project in Zambia.

The opinion of the panel was rather less clear cut with regard to whether local people should also be directly involved in the processes of developing, planning and managing ecotourism. While only 20% of the original set of definitions made explicit reference to this condition, by the end of the second round just over 27% of the definition set included the precondition that ecotourism should actively seek the participation of local people. While this figure has evidently risen over the course of the study, it is nevertheless surprisingly low given the importance accorded to local participation in the literature. Indeed, Paul (1987), Drake (1991) and Brandon (1993) all make strong cases for local participation going beyond simply making locals into economic beneficiaries of ecotourism, involving local people fully in all aspects and at all stages of the development, planning and management of ecotourism. One possible explanation might lie in the nature of the Delphi process, which engages experts already involved in the practice of ecotourism. The positions or influence of these individuals might be felt to be threatened if more local participation were to be encouraged in the planning and management of the projects for which they are responsible (Drake, 1991).

An issue still unresolved in this debate, however, is the extent to which economic benefits are considered to be of value in themselves, or merely as a method for encouraging local people to commit to the conservation goals of ecotourism. It can be argued that ecotourism often represents a viable means of economic diversification, especially for peripheral areas (Garrod *et al.*, 2001). Such areas typically have few other economic resources available on which local people can base their livelihoods. In such circumstances, ecotourism can represent a means of sustainably exploiting the marine environment – an activity in which many peripheral areas have a distinct comparative advantage in that they are able to supply the kind of high-quality marine environment that will be prized by ecotourists. It is often relatively easy for local communities with a maritime heritage to diversify into marine ecotourism, which can directly employ some of the resources previously engaged in the fishing industry or already in place serving the conventional tourism sector. Marine ecotourism can also have significant multiplier effects, making a strong contribution to local income and employment (MacLellan, 1999). Of course, such economic benefits are only going to be genuinely sustainable if ecotourism serves to protect and enhance the natural resource base, rather than to deplete or degrade it. Hence, unlike many other economic activities that have achieved only a temporary increase in the prosperity of the locations to which they are introduced, it is argued that marine ecotourism can be a potential vehicle for the sustainable regeneration of peripheral areas (Garrod *et al.*, 2001).

While the foregoing discussion might be taken to imply that the panel members believed the rationale for ecotourism to be primarily an economic one – emphasising the benefits to local people rather than to nature conservation – such a conclusion cannot be sustained by the data. Indeed, while 60% of the definitions in the original set claimed conservation as a principal rationale for ecotourism, this figure increased to just over 80% by the end of the second round. This would suggest that the predominant view is, in fact, that ecotourism must achieve both ends simultaneously; in other words that it must both benefit local people economically and be an effective tool for conservation.

Another finding of this analysis is that the panel members evidently did not feel that the cultural component of ecotourism was essential to its definition. Cultural features associated with ecotourism, such as local customs, languages, arts and crafts, and food, were explicitly included in some of the earliest and best-known definitions of ecotourism (see, for example, Ceballos-Lascuraín, 1987). Indeed, while 40% of the initial set of definitions included reference to cultural features, only just over 18% of the definition set achieved by the end of the second round included explicit reference to associated cultural features. This might suggest that the panel members preferred to view ecotourism as being based primarily on the natural environment. However this view is not supported by the data on the 'nature-based' node – while 50% of the original set of definitions included reference to ecotourism being nature based, by the end of the second round this figure had fallen to just over 36%. Meanwhile the view that ecotourism should be expected to take place only in relatively undisturbed natural areas was not widely supported by the panel members. While 30% of the original definitions included this criterion, by the end of the second round this figure had fallen to only 9%.

Regarding the length of definitions, the content analysis showed little movement in the distribution of preferences between the three size groups. While 30% of the initial definition set were 30 words or fewer in length, 50% between 31 and 60 words, and the remaining 20% more than 60 words long, by the end of the second round these figures had changed only slightly to around 27%, 54% and 18% respectively. This might suggest that, on the whole, panel members were happiest with medium-length definitions. The reason might be that these are considered short enough to be reasonably clear cut and memorable, yet long enough to capture the various subtleties of what distinguishes 'true' ecotourism from 'counterfeit' ecotourism. Such imposters are typically attempting to capture the marketing benefits from employing the ecotourism 'brand' (which can be considerable) without having to abide by the ethics of ecotourism, which implies adopting measures that might limit the market potential of their 'product' (see Weaver, 1998; Wight, 1993).

Finally, with regard to the preferred style of definition, the findings of the content analysis would suggest that prescriptive rather than descriptive definitions were generally preferred. Definitions that only described what practical form ecotourism might take or what activities it might involve (Styles 1 and 2) were not preferred. While 10% and 20% of the initial definition set were classified as Styles 1 and 2 respectively, by the end of the second round these figures had each fallen to only around 9%. Definitions that were highly prescriptive, sometimes to the extent of setting out criteria for candidate activities to fulfil if they are to be considered examples of ecotourism (Styles 4 and 5), were neither strongly accepted nor rejected. While 20% and 30% of the initial definition set were considered to be written in Styles 4 and 5 respectively, by the end of round these figures had both fallen slightly to around 18% and 27% respectively, denoting little overall change. Definitions occupying the middle ground (Style 3), meanwhile, actually increased in acceptance in the course of the study, with the proportion of definitions in the chosen sets rising from 20% at the beginning of the study to over 36% at the end.

Discussion and conclusions

This chapter has attempted to explore what ecotourism means to the experts who are involved either in setting the agenda for its development, planning and management or directly involved in planning, managing or providing such activities. By exploring what ecotourism means to influential people – what they conceive it to be and how, therefore, they intend to go about achieving it – we may now draw together a number of useful lessons for ecotourism developers, planners and managers more generally.

The main findings of the content analysis suggest that while the key components of the ecotourism concept are widely appreciated – the need for suitable education and interpretation, the need to manage ecotourism appropriately and the aim of genuine sustainability – other aspects remain contested. For example, there would seem to be a significant difference of opinion as to whether ensuring that the economic benefits of ecotourism remain in the hands of local people should be seen as an end in itself, as a means of ensuring that local people commit to the conservation objectives of the ecotourism project or as a combination of both. Similarly, the issue of whether local people should be directly involved in the development, planning and management of ecotourism is emphasised by some yet seen as relatively unimportant by others. Such differences of view are important, since the way in which the crucial issues involved in ecotourism are perceived by decision-makers can have critical implications for the way in which it is carried out in a particular location.

With regard to the definitions themselves, a number of interesting lessons flow from the analysis. Firstly, medium-sized definitions appear to draw the most support from the experts, suggesting that a favourable balance between tightness of definition and content needs to be drawn. Ecotourism is clearly a complex subject and we should not expect to be able to communicate the intricacies of its true meaning in a 'short and snappy' definition. Yet, at the same time, the importance of keeping any proposed definition short enough to be accessible to those it is intended to influence is also highlighted by this analysis. Clearly there is a fine line to be drawn between completeness and depth on the one hand, and tightness and accessibility, on the other.

Second, the content analysis suggests that there is also a fundamental balance to be drawn between description and prescription. While a definition of ecotourism should clearly prescribe the kinds of activity, location and market towards which the ecotourism experience should be oriented, it is important not to limit the definition too narrowly. Indeed, if it is believed that ecotourism is best seen as a process rather than a form of tourism, then it would be unwise to exclude certain instances of tourism simply because they do not fit within one's normal conception of what form ecotourism should take (be it only in 'wilderness' areas, entirely based on viewing wildlife, 'non-consumptive' or whatever). This argument suggests that while strict criteria-based definitions of ecotourism may be attractive to academics, they may be seen as less useful in the practical context of developing, planning and managing ecotourism.

Third, the study also raised some interesting issues with respect to the advantages and disadvantages of defining ecotourism. The main advantages would appear to be that definitions are useful in order to communicate the objectives of ecotourism to local people and as a means of reminding everyone involved in developing, planning and managing ecotourism of the ethical underpinnings of the ecotourism concept. Another widely supported advantage was thought to be that defining ecotourism helps to focus the input of the disparate bodies that are typically involved in ecotourism towards a common aim. On the other hand, the major disadvantage of defining ecotourism was considered to be that too narrow a definition of ecotourism would not allow the concept to be applied in certain circumstances or make certain stakeholder groups feel excluded or marginalised. Another widely supported disadvantage was that as the concept becomes more widely known and fashionable there is a danger that it may be used as a 'green label' for existing or potential operators to use as a marketing tool, cynically exploiting people's goodwill towards the concept. The overall lesson to be learned is that a common definition of ecotourism would be valuable provided that it is used responsibly by those developing, planning and managing such activities.

Acknowledgements

This study was conducted as part of an EU Interreg IIc project (No. EA-C1UK-No.3.14) entitled 'Marine Ecotourism for the Atlantic Area (META-)'. Special thanks are due to Julie Wilson and David Bruce for their collaboration on the project. The gallant efforts of the Delphi panel members in interpreting the questions, completing the questionnaires and meeting the deadlines are also much appreciated.

References

Blamey, R.K. (1995) The nature of ecotourism. Occasional Paper No. 21. Canberra: Bureau of Tourism Research.

Blamey, R.K. (1997) Ecotourism: The search for an operational definition. *Journal of Sustainable Tourism* 5 (2), 109–30.

Brandon, K. (1993) Basic steps toward encouraging local participation in nature tourism projects. In K. Lindberg and D.E. Hawkins (eds) *Ecotourism: A Guide for Local Planners* (pp. 134–51). North Bennington: The Ecotourism Society.

Burton, F. (1998) Can ecotourism objectives be achieved? *Annals of Tourism Research* 25 (3), 755–8.

Cater, E. (1992) Profits from paradise. *Geographical Magazine* 64 (Mar.), 16–21.

Ceballos-Lascuraín, H. (1987) The future of ecotourism. *Mexico Journal* (Jan.), 13–14.

Commonwealth Department of Tourism (1992) *Australian National Ecotourism Strategy*. Canberra: Commonwealth Department of Tourism.

Dalkey, N. and Helmer, O. (1963) An experimental application of the Delphi method to the use of experts. *Management Science* 9 (3), 458–67.

Diamantis, D. (1999) The concept of ecotourism: Evolution and trends. *Current Issues in Tourism* 2 (2&3), 93–122.

Drake, S.P. (1991) Local participation in ecotourism projects. In T. Whelan (ed.) *Nature Tourism: Managing for the Environment* (pp. 132–63). Washington, DC: Island Press.

Ecotourism Society (1994) On WWW at http://www.ecotourism.org.

Evans-Pritchard, D. and Salazar, S. (1992) *What is Ecotourism?* Costa Rica: Eco Institute of Costa Rica and ULACIT.

Fennell, D.A. (1999) *Ecotourism: An Introduction*. London: Routledge.

Garrod, B. (2002) Preface. *International Journal of Sustainable Development* 5 (3), 227–31.

Garrod, B., and Fyall, A. (2000) Managing heritage tourism. *Annals of Tourism Research* 27 (3), 682–708.

Garrod, B., Wilson, J. and Bruce, D. (2001) *Planning for Marine Ecotourism in the EU Atlantic Area: Good Practice Guidance*. Bristol: University of the West of England.

Green, C., Hunter, C. and Moore, B. (1990) Assessing the environmental impact of tourism development: Use of the Delphi technique. *Tourism Management* 11 (2), 111–20.

Helmer, O. (1972) *On the Future State of the Union*. Report 12–2. Menlo Park, CA: Institute for the Future.

Hill, K.Q. and Fowles, J. (1975) The methodological worth of the Delphi technique. *Technological Forecasting and Social Change* 7 (2), 179–92.

Inskeep, E. (1999) *Guide for Local Authorities on Developing Sustainable Tourism: Supplementary Volume on Sub-Saharan Africa*. Madrid: World Tourism Organisation.

Kaynak, E. and Macaulay, J.A. (1984) The Delphi technique in the measurement of tourism market potential: The case of Nova Scotia. *Tourism Management* 5 (2), 87–101.

Linstone, H.L. and Turoff, M. (eds) (1975) *The Delphi Method: Techniques and Applications*. Reading, MA: Addison-Wesley.

Liu, J.C. (1988) Hawaii tourism to the Year 2000: A Delphi forecast. *Tourism Management* 9 (4), 279–90.

MacLellan, L.R. (1999) An examination of wildlife tourism as a sustainable form of tourism development in North West Scotland. *International Journal of Tourism Research* 1 (5), 375–87.
Masser, I. and Foley, P. (1987) Delphi revisited: Expert opinion in urban analysis. *Urban Studies* 24 (2), 217–25.
Miller, G. (2001) The development of indicators for sustainable tourism: Results of a Delphi survey of tourism researchers. *Tourism Management* 22 (4), 351–61.
Orams, M.B. (1995) Towards a more desirable form of ecotourism. *Tourism Management* 16 (1), 3–8.
Page, S.J. and Dowling, R.K. (2002) *Ecotourism*. Harlow: Pearson Education.
Pan, S.Q., Vega, M., Vella, A.J., Archer, B.H. and Parlett, G. (1995) A mini-Delphi approach: An improvement on single round techniques. *Progress in Tourism and Hospitality Research* 2 (1), 27–39.
Paul, S. (1987) *Community Participation in Development Projects: The World Bank Experience*. Washington, DC: World Bank.
Richey, J.S., Mar, B.W. and Horner, R.R. (1985) The Delphi technique in environmental assessment I: Implementation and effectiveness. *Journal of Environmental Management* 21 (1), 135–46.
Ross, S. and Wall, G. (1999) Evaluating ecotourism: The case of North Sulawesi, Indonesia, *Tourism Management* 20 (6), 673–82.
Rotundi, A. and Gustafson, D. (1996) Theoretical, methodological and practical issues arising out of the Delphi method. In M. Adler and E. Zigler (eds) *Gazing into the Oracle: The Delphi Method and its Application to Social Policy and Public Health* (pp. 34–55). London: Jessica Kingsley.
Sackman, H. (1975) *Delphi Critique*. Lexington: DC Heath and Company.
Smith, S.L.J. (1995) *Tourism Analysis: A Handbook* (2nd edn). Harlow: Longman.
Teeling-Smith, G. (1971) Medicines in the 1990's: Experience with a Delphi forecast. *Long Range Planning* 3 (4), 69–74.
Tremblay, P. (2001) Wildlife tourism consumption: Consumptive or non-consumptive? *International Journal of Tourism Research* 3 (1), 215–18.
Valentine, P.S. (1993) Ecotourism and nature conservation: A definition with some recent developments in Micronesia. *Tourism Management* 14 (2), 107–15.
Wall, G. (1997) Is ecotourism sustainable? *Environmental Management* 21 (4), 483–91.
Weaver, D. (1998) *Ecotourism in the Less Developed World*. Wallingford: CAB International.
Weaver, D. (1999) Magnitude of ecotourism in Costa Rica and Kenya. *Annals of Tourism Research* 26 (4), 792–816.
Wheeller, B., Hart, T. and Whysall, P. (1990) Application of the Delphi technique: A reply to Green, Hunter and Moore. *Tourism Management* 11 (2), 121–2.
Wight, P. (1993) Ecotourism: Ethics or eco-Sell? *Journal of Travel Research* 31 (3), 3–9.

Chapter 2
Between the Devil and the Deep Blue Sea: Dilemmas for Marine Ecotourism

ERLET CATER

Introduction

If outer space is the 'final frontier', it may be argued that the marine environment is the penultimate, with advanced technology enabling an increasing number of marine tourists literally to reach new depths and to view marine life in its natural habitat without even getting wet. The Underwater Observatory in Milford Sound, New Zealand, the semi-submersible Nessee in Mauritius, the glass-hulled Seaprobe Atlantis in Kyle of Lochalsh, Scotland, and the tourist submarine SADKO in Larnaca, Cyprus, are all examples of facilities that enable the observation of marine ecosystems previously confined to those who were scuba-qualified.

Marine tourism is one of the fastest growing tourism market segments (Orams, 1999). The last decade has witnessed a global proliferation of marine tour operators offering experiences ranging from boat trips to view marine life in the Inner Hebrides to undersea walks at Grand Baie, Mauritius. Whalewatching experiences are now on offer at approximately 500 locations around the world. Over 9 million participants were recorded in 1999 (see Figure 2.1), displaying an average annual percentage growth rate worldwide of 12.1% throughout the 1990s (Hoyt, 2001). Also in 1999, the world's largest diving organisation, the Professional Association of Dive Instructors (PADI), issued in excess of 800,000 new certifications (see Figure 2.2), accounting for a total of over 9 million since 1967 (PADI, 2001).

The central dilemma, however, and hence the title of this chapter, is how to manage such activity sustainably, so that it can lay a justifiable claim to the title of marine ecotourism.

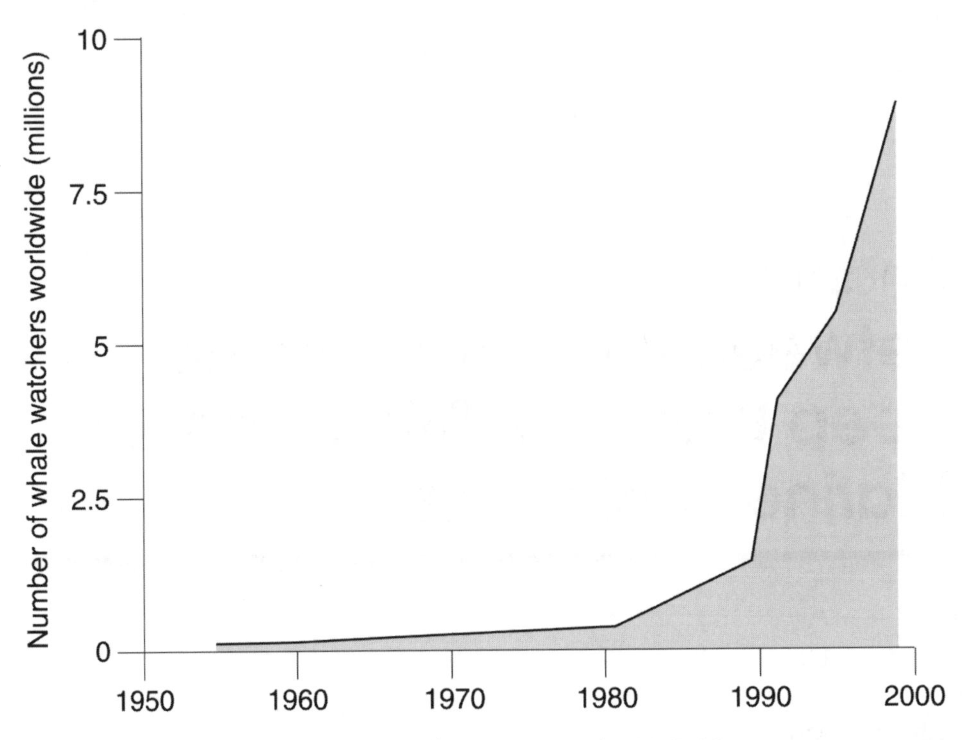

Figure 2.1 The growth of whalewatching worldwide, 1955–98
(*Source*: Hoyt, 2001)

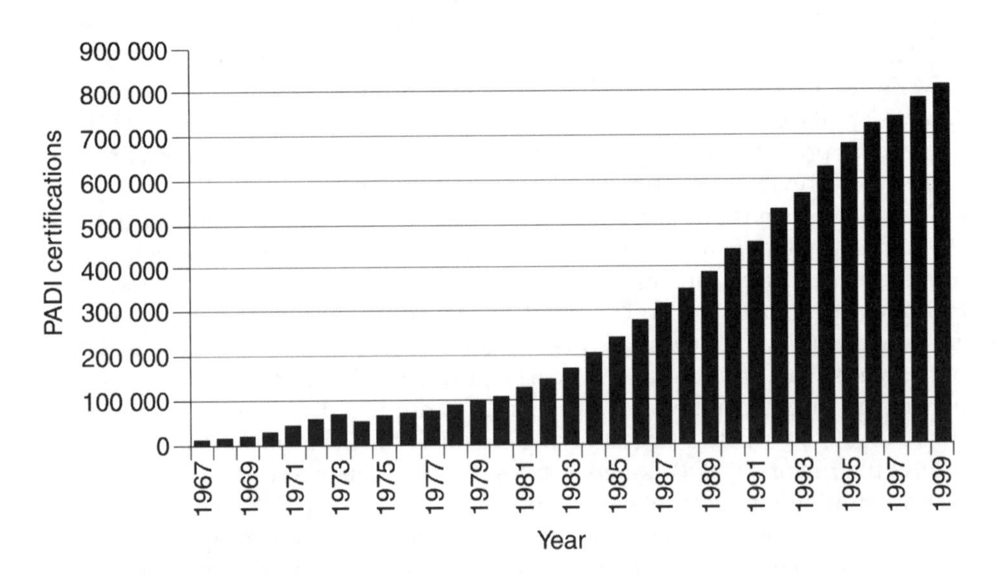

Figure 2.2 PADI certifications history worldwide: all levels 1967–99

Sustainable Marine Ecotourism in Theory

As with any form of nature-based tourism, it is essential that marine ecotourism should be sustainable, incorporating the qualities of socio-cultural responsibility, consumer satisfaction, economic viability and environmental integrity. If marine ecotourism embodies these principles, symbiotic relationships between the varying interests should follow, with marine environmental protection resulting both *from* and *in* enhanced coastal livelihoods, continued profits for the tourism industry, sustained visitor attraction and revenue for marine conservation (see Figure 2.3). Theoretically speaking, this is a non-destructive operationalisation of the 'use it or lose it' philosophy: simultaneously making conservation pay and paying for conservation. So much for the theory, but what about the practice?

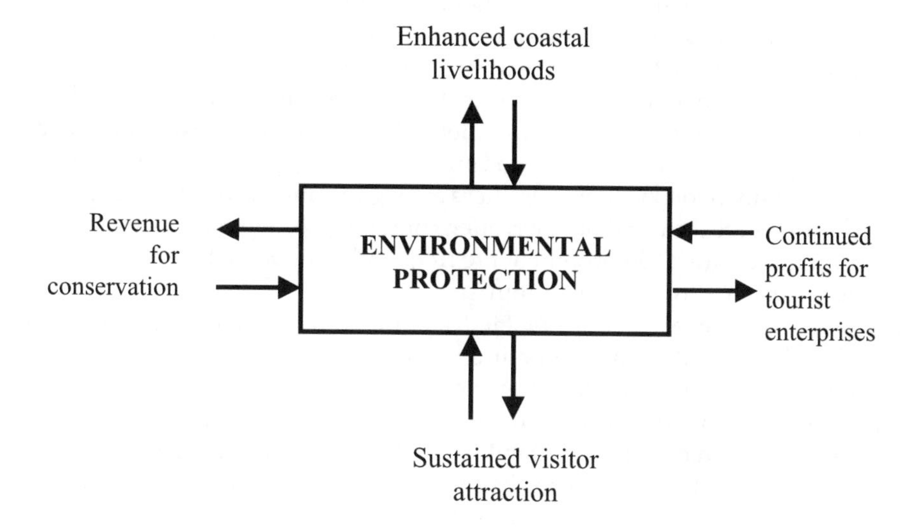

Figure 2.3 Symbiotic relationships in marine ecotourism

Sustainable Marine Ecotourism in Practice

The gap between theory and practice

The potential benefits of marine ecotourism are widely documented but despite various guidelines, such as those of the International Ecotourism Society (Halpenny, 2002), there is a considerable gap between theory and practice. The disappointing performance of marine ecotourism to date in achieving several of its goals, such as increased local involvement, indicates the need to identify not only the nature of this gap but also its causes. One cause is the failure to recognise the wider context within which marine ecotourism is cast as a process and a principle.

The wider context

There is an inherent danger of merely regarding marine ecotourism as an isolated alternative, disassociating it from all other forms of economic activity. The context in which it is set as a process and as a principle is all important, because that context has a vital role to play in prospects for sustainable outcomes (Cater & Cater, 2001). Other activities may 'make or break' ecotourism. There are also vital ramifications for the success or failure of development strategies that place a central emphasis on the contribution of ecotourism towards enhanced livelihoods in developing countries. Donor agencies are becoming increasingly interested in ecotourism as a pro-poor strategy (Ashley *et al.*, 2001) but unless it is recognised that it is but one component of such a strategy – and frequently a minor one at that – then the exercise is likely to be doomed to failure. At most, it should be viewed as a complementary or supplementary activity – not a substitute.

Recognition of the wider context involves understanding the very important two-way relationships, which operate at various scale levels (see Figure 2.4), with the positive, or spread, effects of marine ecotourism ideally diffusing through the hierarchy. These include raising environmental awareness and disseminating an understanding of the coincidence of good environmental practice with advantages to business. Simultaneously, however, there are significant backwash, or negative, effects. These include the fact that other, often competing, economic activities which impinge upon the natural environment are frequently prejudicial to the success, if not the very existence, of marine ecotourism.

There are various levels to consider. First, it is imperative that marine ecotourism is viewed in the context of marine nature tourism as a whole, as the activities of unprincipled nature tour operators may compromise genuine marine ecotourism. Such environmental opportunists unjustifiably, and misleadingly, hijack the prefix 'eco' to confer an image of respectability to their operations, for example the shark-cage diving activities at Gansbaii, Cape Province, South Africa, or the 'eco-voyeuristic' whalewatching experiences advertised in the Dominican Republic (DR pure, 2001). Such businesses may be ecologically based but not ecologically sound, socio-culturally responsible or even, if any of these qualities are compromised, ultimately economically viable. Conscientious marine ecotourism operators may find their efforts constantly thwarted by the unsustainable activities of other marine nature tour operators. An example is the award-winning sea kayaking operations of SeaCanoe in Thailand. The very success of SeaCanoe in an unregulated situation inevitably spawned unscrupulous imitators. Sea cave visitation has been taken into four figures a day, together with the illegal extortion of tourist entry fees. Inevitably the caves have become degraded by these high volume, environmentally destructive entries (Gray, 1998a, b). Independent visitation also brings its problems. There is considerable concern over the decline of numbers of dolphins at Brixham, Devon, which has largely been attributed to harassment by careless small boat users (Cornwall Wildlife Trust, 2001).

Second, marine ecotourism needs to be considered with respect to other tourism market segments which are dependent upon, and consequently impact on, the

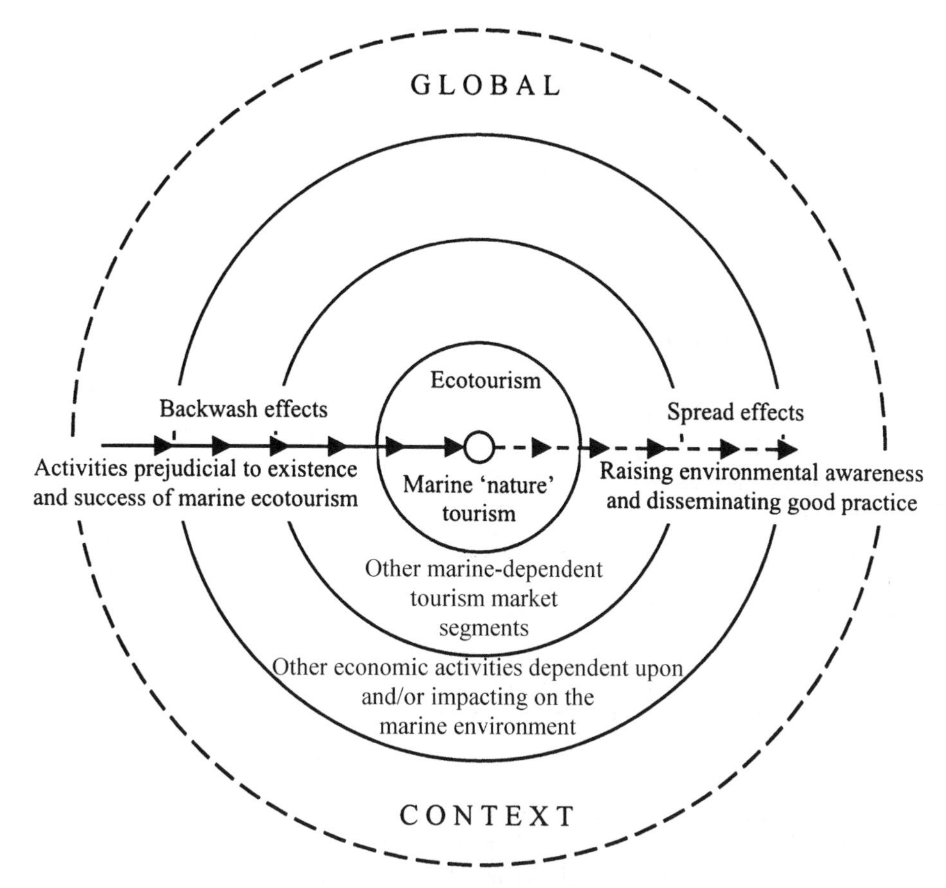

Figure 2.4 Spread and backwash effects of marine ecotourism

marine environment. The development and operations of coastal resorts, for example, have manifest implications for the success or otherwise of marine ecotourism. The construction of 120 hotels in 20 years at the mega-resort of Cancun, Mexico, has endangered breeding areas for marine turtles, as well as causing large numbers of fish and shellfish to be seriously depleted just offshore (Mandala Project, 2001).

Any one location is likely to host a variety of (frequently incompatible) recreational pursuits. There is a definite danger that the requirements of one activity will prejudice the needs of another, for example, the conflict between scuba-diving and high-speed watercraft. It can be seen, therefore, that it is vital that marine ecotourism is viewed not only in the context of marine nature-based tourism but also in that of tourism in general. As Milne (1998: 47) suggests: 'in attempting to achieve more appropriate forms of tourism, it is also essential that we steer away from creating a dichotomy between "alternative" and "mass" tourism. Such a

Table 2.1 Sectors that may have a stake in, or impact upon, marine ecotourism

Sectors
Agriculture
Coastal & ocean research
Dredge and spoil disposal
Fisheries
Forestry
Housing
Mariculture / aquaculture
Marine industry & power production
Military areas and facilities
Ocean engineering and technology
Oil & gas extraction
Ports / harbours / marinas
Protected areas
Shipping & navigation
Solid & hazardous waste disposal
Water pollution / pollution control
Water supply
Wildlife management & nature conservation

Source: Garrod *et al.* (2001).

division serves little real purpose and diverts our attention away from the interlinked nature of all types of tourism development'.

Third, with regard to the overall picture of sustainability, it is vital to consider the interactions that occur between marine ecotourism and all other forms of activity which impinge on the marine environment. It is vital that a move is made beyond a tourism-centric view, as it is 'inappropriate to discuss sustainable tourism any more than one might discuss any other single activity ... we cannot hope to achieve sustainability in one sector alone, when each is linked to and dependent upon the others' (Butler, 1998: 28). Sectoral conflicts may compromise the success, if not the very existence of ecotourism. Garrod *et al.* (2001a) list the many sectors that may have a stake in, or impact upon, marine ecotourism (see Table 2.1). A number of these constitute serious threats. The effects of destructive fishing practices such as the use of dynamite and cyanide on coral reef ecosystems in Indonesia and the Philippines, for example, have been well documented (White, 1986; Wells & Hanna, 1992). Toxic spillages from ocean transport, such as the oil spillage in the Galapagos in January 2001 or that of rat poison off Kaikoura, New Zealand in June 2001, which threatened the world renowned operations of Whale Watch (Whale and Dolphin Conservation Society, 2001), are also omnipresent threats. Run-off from land-based activities also has manifest implications in terms of pollution, siltation and eutrophication. Destructive logging practices in Southeast Asia result in considerable coastal siltation, posing a threat to coral reefs. In the UK, it is interesting to note that timber extracted from remote Scottish West Highland and Island forests is being increasingly removed by sea transport. The Forestry Commission estimates that 750,000 tonnes of timber, worth £1.5 million, is scheduled to be transported by such means from otherwise inaccessible locations in north-west Scotland (West Highland Free Press, 10 August 2001). Such activity is bound to detract from the visual quality of a pristine setting, as indeed does the proliferation of salmon cages in beautiful sea lochs. Considerable concern is also voiced about the environmental effects of fish farming. The extent of nutrient pollution from Scottish aquaculture in 2000 was estimated to be 7500 tonnes of nitrogen, comparable to the annual sewage inputs of 3.2 million people; and 1240 tonnes of phosphorus, comparable to that from 9.2 million people. This is likely to affect important Highland and Island marine habitats such as seaweed forests, where resultant cloudiness of the water reduces the depth to which forests can grow. There is also a major economic threat posed by the proliferation of toxic blooms attributable to the distortion of nutrient ratios. Apart from commercial fishery implications, an increase in toxic blooms has wider implications for wildlife as they can be lethal to birds and sea mammals, for which Scotland has a significant international reputation (Berry & Davison, 2001). Another contentious issue is that of military activity. The harmful effects of the testing of naval underwater sonar systems on marine mammals has been recorded in several locations, including Greece, the Bahamas and Hawaii (Lazaroff, 2000; NRDC, 2001), and remain an unknown quantity in the deeper waters of West Scotland where such activities also take place. Parsons *et al.* (2000) present evidence that military sonar uses frequencies to which cetaceans occurring in the Hebrides would be sensitive.

The significance of extractive activities to coastal livelihoods must not be denied. Fish farming, for example, offers year-round employment to remote coastal communities with few other sources of income but the need for sustainable practices is clear. If other activities result in the destruction of the resources upon which marine ecotourism depends, its very existence, let alone success, will be placed in jeopardy.

Finally, it is essential to consider the global context within which marine ecotourism is cast as a process. Tourism is an industry with a global reach, tourists search for ever more different and varied experiences. It is no coincidence, therefore, that marine nature tourism is experiencing growth rates far in excess of tourism as a whole. Also, in an increasingly interdependent world, the interconnected world becomes one which is easily sabotaged (Johnston *et al.*, 1995). Marine ecotourism attractions are often in remote, relatively poorly policed, locations where tourists become pawns in the power struggles of factional groups headquartered away from the seat of government. The kidnapping by the Philippine Abu Sayyaf movement of 21 hostages, including 10 dive tourists, from the Malaysian island of Sipadan in April 2000, and the abduction of 20 tourists from the Philippine island of Palawan in May 2001, are cases in point. The global environmental context is also important: the implications of global climate change for coral bleaching and consequent destruction of a significant marine ecotourism resource is an obvious manifestation of this fact (Brown, 1997). Basking sharks were recorded around the headlands of Devon, weeks earlier than usual, in May 2001 (Cornwall Wildlife Trust, 2001; see also Speedie, this volume).

It is evident from this discussion that it is vitally important that marine ecotourism is not considered in isolation: it cannot stand alone, and it is not a universal panacea for unsustainable practices. Circumstances will dictate its success or failure in terms of sustainable outcomes. These, of course, will be place-specific but the message for collaboration is clear, both within the activity (intra-sectoral) and between different sectors and interests (inter-sectoral). As Bramwell and Lane (2000: 4) describe, collaboration 'involves relationships between stakeholders when those parties interact with each other in relation to a common issue or "problem domain"'. They go on to point out that resource dependency and stakeholder interdependence mean that there are potential mutual or collective benefits arising from stakeholders collaborating with each other.

Intra and Inter-sectoral Collaboration

In terms of *intra*-sectoral collaboration, a variety of new marine tourism operator networks are emerging. These can perform important roles in marketing (frequently marine ecotourism businesses are small and isolated, and lack the financial capital to reach the marketplace effectively); in developing codes of practice; and in having a collective voice to influence policy. The Skye and Lochalsh Marine Tourism Association (SLMTA) was established 'to provide members with a mechanism for sharing good practice and to promote members' services to interested parties' (SLMTA, 2001: 1). The Scottish Marine Wildlife Operators Association (SMWOA), formed in 1998, admits marine wildlife tourism operators who agree to adopt their code of practice but also acts as a point of contact with the Scottish Tourist Board as well as voicing members concerns to statutory agencies as well as conservation groups (SMWOA, 2001).

Inter- or *cross-sectoral* partnerships are 'engaged in developing policies and planning that go beyond basic tourism questions: they also deal with broader economic, social and environmental issues' (Bramwell & Lane, 2000: 4). According to Bramwell and Lane, there are four main ways in which such collaborative approaches should help further sustainable development:

> First, collaboration among a range of stakeholders including non-economic interests might promote more consideration of the varied natural, built and human resources that need to be sustained for future well-being. Second, by involving stakeholders from several fields of activity, with many interests, there may be greater potential for the integrative or holistic approaches to policy- making that can help to promote sustainability ... Partnerships can also help reflect and help safeguard the interdependence that exists between tourism and other activities and policy fields ... Third, if multiple stakeholders affected by tourism development were involved in the policy-making process, then this might lead to a more equitable distribution of the resulting

> benefits and costs. Participation should raise awareness of tourism impacts on all stakeholders, and this heightened awareness should lead to policies that are fairer in their outcomes. Fourth, broad participation in policy-making could help democratise decision-making, empower participants and lead to capacity building and skill acquisition amongst participants and those whom they represent. (Bramwell & Lane, 2000: 4)

One framework which considerably facilitates integration of all sectors, levels and interests is that of Integrated Coastal Zone Management (ICZM).

ICZM is defined as 'a process which brings together all those involved in the development, management, and use of the coast within a framework which facilitates the integration of their interests and responsibilities to achieve common objectives' (UK DoE, cited in Garrod *et al.*, 2001b). It is being developed in many parts of the world to maximise the benefits provided by the coastal zone and to minimise the conflicts and harmful effects of activities upon each other, on resources and on the environment. Between 1996 and 1999, the European Commission operated a Demonstration Programme on ICZM designed around a series of 35 demonstration projects and six thematic studies. In 2000, the European Commission adopted two documents based on the experiences and outputs of the Demonstration Programme and it was the intention that the European Council's common position should be adopted at the end of October 2001 (ENV ICZM, 2001).

Of course, collaboration is not without its problems: one of the most vaunted principles of ecotourism is that of local involvement. However, we are often talking about a set of very unequal power relations. Jamal and Getz (2000) highlight how stakeholders may not only have a variety of needs and desires but also have differing abilities to influence the agenda and scope of investigations. They voice a concern that the interests of 'other', less affluent, less visible segments of the community might not be considered adequately or indeed that they are even kept informed about the process. This raises the question of empowerment and capacity raising: goals which are not achieved overnight. On Apo island, in the Philippines, the Marine Conservation and Development Program was formally implemented in 1984 following five years of preliminary activities. One of the prime objectives of this programme was to establish a community group that would take charge of the conservation and management of the island's coastal resources. Such local 'ownership', as opposed to imposition from outside, has helped to secure enduring sustainable outcomes (M. Pascobello, Apo, 2001, pers. comm.). One of the lessons learned from Apo, however, was that community-based coastal resource management is a long and never-ending undertaking (Calumpong, 2000), which requires sustained commitment. In the UK, the objectives of the proposed Comann na Mara (Society of the Sea) at Loch Maddy, North Uist, Scotland, include the development of marine science and services for fishermen, fish farmers and other sea users. Central to Comann na Mara's objective of fostering sustainable development of the marine environment by encouraging its sensitive stewardship is that not only should its proposed marine interpretation centre constitute a visitor attraction, but, more fundamentally, it should act as a 'drop-in' local resource centre for local fishermen as well as performing a catalytic role for visiting students and marine scientists (J. McLeod, North Uist, 2002, pers. comm.). In March 1999, the Seaprobe Atlantis was chartered for a special exercise in community education by the Loch Maddy Marine Special Area of Conservation (SAC) and 281 local residents were taken on half-hour trips to view the underwater ecology of this sea loch (A. Rodger, North Uist, 1999, pers. comm.).

So, it can be seen that marine ecotourism, in many instances, may indeed be caught between the 'devil', in the form of negative, backwash effects from a multiplicity of interacting sectors, levels and interests, and the 'deep blue sea', resulting from the desirable, positive spread effects which have sustainable outcomes. The overall challenge is for collaborative efforts to be made in order to understand areas of discord, as well as concord, so that negative links can be broken and positive links built upon (World Bank, 1992; Cater, 1995). Only then will marine ecotourism across the globe begin to live up to the reputation of sustainability that goes before it. There is an increased awareness of how much the human population depends upon a healthy marine environment for sustainable livelihoods. The quality of these livelihoods is also increasingly dependent on leisure opportunities that allow us to appreciate, understand and thus help to safeguard the remarkable diversity of marine environments.

References

Ashley, C., Goodwin, H. and Roe, D. (2001) Pro-poor tourism strategies: Expanding opportunities for the poor. Pro-poor Tourism Briefing, No. 1. April. London: ODI, IIED and CRT.

Berry, C. and Davison, A. (2001) Bitter harvest: A call for reform in Scottish aquaculture. Report for WWF Scotland, Perthshire.

Bramwell, B. and Lane, B. (2000) Collaboration and partnerships in tourism planning. In B. Bramwell and B. Lane (eds) *Tourism Collaboration and Partnerships: Politics, Practice and Sustainability* (pp. 1–19). Clevedon: Channel View.

Brown, B. (1997) Coral bleaching: Causes and consequences. *Coral Reefs* 16, 129–38

Butler, R. (1998) Sustainable tourism: Looking backwards in order to progress? In C.M. Hall and A.A. Lew (eds) *Sustainable Tourism: A Geographical Perspective* (pp. 25–34). Harlow: Longman.

Calumpong, H. (2000) *Community Based Coastal Resources Management in Apo Island.* Silliman University, Philippines.

Cater, E. (1995) Environmental contradictions in sustainable tourism. *Geographical Journal* 161 (1), 21–8.

Cater, C. and Cater, E. (2001) Marine Environments. In D. Weaver (ed.) *The Encyclopaedia of Ecotourism.* Wallingford: CABI.

Cornwall Wildlife Trust (2001) Don't distress the dolphins (online). 22 May [http://www.wildlifetrust.org.uk/cornwall/news/devondol.html].

DR pure (2001) Whalewatching: Adventure tourism in the Dominican Republic (online). DR Pure, Dominican Republic 10 June [http://drpure.com/drpure/index.php?topic = Whale].

ENV ICZM (2001) *Integrated Coastal Zone Management* (online). 14 August [http://europa.eu.int/comm/environment/iczm/home/html].

Garrod, B., Wilson, J. and Bruce, D. (2001a) *Planning for Marine Ecotourism in the EU Atlantic Area: Good Practice Guidance.* Bristol: University of the West of England.

Garrod, B., Wilson, J. and Bruce, D. (2001b) International and EU obligations and duties for planners of marine ecotourism: An informal summary (online). 14 August [www.tourism-research.org/obligs.doc].

Gray, J. (1998a) SeaCanoe Thailand: Lessons and observations. In M. Miller and J. Auyong (eds) *Proceedings of the 1996 World Congress on Coastal and Marine Tourism.* University of Washington, Seattle, and Oregon Sea Grant Program (pp. 139–44). Oregon State University.

Gray, J. (1998b) Update on SeaCanoe wars. *Trinet* (newsgroup) 2 December [caveman@seacanoe.com].

Halpenny, E. (2002) *Marine Ecotourism: Impacts, International Guidelines and Best Practice Case Studies*. Burlington: International Ecotourism Society.

Hoyt, E. (2001) Whalewatching 2001: Worldwide tourism numbers, expenditures, and expanding socioeconomic benefits. Special Report from the International Fund for Animal Welfare. Yarmouth Port: IFAW.

Jamal, T. and Getz, D. (2000) Community roundtables for tourism-related conflicts: The dialectics of consensus and process structures. In B. Bramwell and B. Lane (eds) *Tourism Collaboration and Partnerships: Politics, Practice and Sustainability* (pp. 159–82). Clevedon: Channel View.

Johnston, R.J., Taylor, P.J. and Watts, M.J. (1995) *Geographies of Global Change.* Oxford: Blackwell.

Lazaroff, C. (2000) Lawsuit claims navy sonar system endangers whales, humans (online). 17 March [http://ens.lycos.com/ens/mar2000/2000L–03–17–07.html].

Mandala Project (2001) Cancun case study: 86 (online). 28 August [http://www.american.edu/TED/index.htm].

Milne, S.S. (1998) Tourism and sustainable development: The global–local nexus. In C.M. Hall and A.A. Lew (eds) *Sustainable Tourism: a Geographical Perspective* (pp. 35–48). Harlow: Longman.

National Resources Defense Council (2001) Navy sonar system threatens marine mammals (online). 20 August [http://www.nrdc.org/wildlife/marine/nfla.asp].

Orams, M. (1999) *Marine Tourism: Developments, Impact and Management.* London: Routledge.

Parsons, E.C.M., Birks, I., Evans, P.G.H., Gordon, J.C.D., Shrimpton, J.H. and Pooley, S. (2000) The possible impacts of military activity on cetaceans in West Scotland. *European Research on Cetaceans* 14, 185–91.

Professional Association of Dive Instructors (PADI) (2001) PADI Certifications History Worldwide (online). 20 June 2001 [http:\\www.padi.com/news/stats/cert_history.asp].

SLMTA (2001) Skye and Lochalsh Marine Tourism Association (online). 17 August [http:\\www.slmta.co.uk].

SMWOA (2001) Scottish Marine Wildlife Operators Association (online). 17 August [http://www.merger.demon.co.uk/smwoa/aboutus.html].

Whale and Dolphin Conservation Society (2001) New Zealand: No wildlife killed in Kaikoura spill (online). 19 July [http://www.wdcs.org/dan/news.nsf/webnews].

Wells, S. and Hanna, N. (1992) *The Greenpeace Book of Coral Reefs.* London: Blandford.

White, A.T. (1986) Marine reserves: How effective as management strategies for Philippine, Indonesian and Malaysian Coral Reef Environments? *Ocean Management* 10, 137–59.

World Bank (1992) *World Development Report 1992.* Oxford: Oxford University Press.

Chapter 3
Planning Policy Issues for Marine Ecotourism

JULIE C. WILSON

Introduction

In its ideal form, marine ecotourism involves a symbiotic relationship between tourism and the natural marine and coastal environment. As a subset of ecotourism, it has considerable global significance in terms of its economic, social, cultural and environmental impacts. Surprisingly, however, specific and formal provisions for the planning and management of marine ecotourism are, in practice, either non-existent or only adopted sporadically and at the most basic level. Furthermore, it is not safe to assume that any tourism activity falling within the general parameters of marine ecotourism automatically equates to sustainable development. Certainly, there are many examples around the world of activities that are portrayed as marine ecotourism but do not meet with the minimum requirements of sustainability. As such, a particular challenge lies in addressing the planning problems that stand in the path of genuinely sustainable marine ecotourism operations.

This chapter introduces some of the problems inherent in the process of planning for marine ecotourism activities. First of all, the actual (and potential) roles of planning for marine ecotourism activities worldwide are examined. Various problems central to planning for marine ecotourism are then evaluated. The chapter then identifies some of the ways in which these problems have been, or could be, addressed, along with some practical examples. Finally, the chapter concludes with some overall considerations for planning in the context of marine ecotourism.

In the first instance, however, a word is required on the notion of 'genuine sustainability', as it pertains specifically to marine ecotourism. It has often been argued that sustainability is the core prerequisite for any ecotourism (see Garrod, this volume), although the term 'genuine sustainability' could arguably be viewed as contentious in itself. Indeed, this basic requirement of sustainability can often represent the weak link between the principles and practices of ecotourism (Garrod, 2002). In reality, while those responsible for ecotourism activity may eagerly sign up to the notion of sustainability, ecotourism has often been driven by the anticipated economic returns. This tends to imply progressive increases in visitor numbers, leading to greater damaging impacts to the natural environment and adverse socio-cultural impacts on the local population.

It is ironic, therefore, that unplanned or poorly planned marine ecotourism can contribute to its own demise, given that genuinely sustainable ecotourism requires the continued provision of a high-quality natural environment. The role of planning within marine ecotourism can, therefore, be seen as crucial to the working out of the principles of ecotourism in practice and this is the subject of the following section.

The Role of Planning in Marine Ecotourism Activity

Planning is a process by which policy is implemented, and a 'plan' is a document that articulates this intended process (Fennell *et al.*, 2001). There are numerous governmental programmes, functions and conditions which are external to [marine] ecotourism but which have significant effects on it (Parker, 2001). Planning is one of these external, yet integral, functions. In ideal terms, statutory planning systems – however useful they may be in their present form – should represent an inextricable part of the dynamic system of marine ecotourism, and this will continue to be the case as marine ecotourism activities increase worldwide. However, whether or not planning processes are fulfilling this kind of role in marine ecotourism is another question entirely. Directly or indirectly, the actual implementation of planning policies worldwide is impeded by the reality that ecotourism (and particularly marine ecotourism) is relatively new and hence relatively weak in terms of its influence (Fennell *et al.*, 2001).

For the most part, statutory and equivalent planning systems are designed to identify and mediate the inevitable conflicts of interest that can arise over the development of land (and offshore). In turn, planning in the context of tourism development processes is often a direct response to the unwanted side effects, particularly at a local level. While it should be recognised that planning is not a panacea, it may have the potential to minimise negative impacts, maximise economic returns to the destination and elicit a more positive long-term response from the host community towards the presence of ecotourism.

In the context of marine ecotourism, planning systems could indeed be used to help resolve conflicts in interests. This applies not just in land use and marine resource conflict terms, but also to the reconciliation of economic opportunities with non-damaging new development (in that planning systems can act as a brake on certain forms of development, if they do not meet specified criteria). As well as conflict reduction benefits, integrated planning can help to avoid replication and increases synergy between development initiatives and can reconcile the different uses of an area for optimal benefit. Planning also has an important role to play in ensuring that local livelihoods are enhanced through the development of marine ecotourism activities which, in turn, can strengthen and support its role in protecting coastal and marine environments. Furthermore, the planning system has a critical potential role in promoting, organising and regulating marine ecotourism activity.

Careful decision-making in planning, where marine ecotourism infrastructure is concerned, can also act to modify or ameliorate other areas of negative impact. Policy and planning may be employed strategically to ensure that the scale of the infrastructure and the scale of tourism are compatible. By way of an example, the case of the Monteverde Cloud Forest Preserve in Costa Rica illustrates how infrastructure decisions either support or impede genuinely sustainable marine ecotourism. The local community within the Preserve made a decision to keep infrastructure improvements simple, specifically by not paving the 40 km dirt access road from the main highway – the Interamericana (as a paved road might well generate more tourists but with less per-person spending as a consequence). If Monteverde could be reached in 30 minutes instead of the usual two and a half hours, then many overnight visitors would instead become day users. In addition, by dint of not staying in the local lodges etc., visitors would generate less income for the residents, who in turn would have a reduced monetary interest in conservation (Honey, 1994, in Parker, 2001: 513).

Unfortunately, this kind of planning intervention has often emerged in an informal context, driven forward by local initiative and foresight. Certainly, statutory planning systems have not fully embraced the responsibility for marine ecotourism planning measures. This may be at least in part due to an overall lack of recognition of marine ecotourism as a coherent form of tourism activity. Indeed, it may be the case that as destination areas that have traditionally relied on other forms of tourism for income, then any formal tourism planning processes have been established and oriented as such. Nonetheless, while there may well be observable weaknesses in current marine ecotourism-relevant planning policy and practice at a global scale, formal planning systems ought to provide the unavoidable and all-important starting point for regulating and promoting marine ecotourism development.

Arguably then, marine ecotourism activities are particularly susceptible to a series of direct problems. However, while it may not be realistic to address fully such inherent weaknesses and problems by changing the statutory mechanisms at an international scale, nor would it be realistic to avoid addressing them altogether. In addition, as there is no particular argument for the introduction of integrated management in such planning-related issues, as would be the case with Integrated Coastal Zone Management (ICZM), subsidiary measures may be necessary. There is, therefore, a major planning challenge for marine ecotourism in alleviating the problems stemming from such weaknesses. Given the potential difficulties in initiating widespread change in regulation and policy at national and international levels, it could be argued that a careful blend of statutory and voluntary structures can alleviate the messy and complex nature of the marine ecotourism planning problem domain. The particular weaknesses and problems that are faced by authorities, local policy-makers and initiators in marine ecotourism planning will now be examined.

Problems in Planning for Marine Ecotourism

Marine ecotourism presents very different kinds of problems to planners and managers, depending on where in the world it is taking place and particularly depending on the local conditions. In the European context, the problems inherent in planning for marine ecotourism appear to parallel the experiences of the EU in ICZM. The EU's ICZM Demonstration Programme observed that many of the physical problems and conflicts observed could be traced back to overall procedural, planning, policy and institutional weaknesses (European Commission, 1999). These weaknesses may indeed extend to marine ecotourism and are, for the most part, connected to a lack of awareness. This relates to a lack of appreciation of the strategic economic and social importance of the sustainable management of tourism in marine and coastal environments (META-Project, 2000).

In an historical context, marine and coastal areas have faced several major planning issues. Marine ecotourism policy and legislative conflicts are often exacerbated by a more general lack of a transnational, inter- and intra-sectoral cooperation which tends to be inherent in many marine issues more generally. However, marine ecotourism activities may be particularly problematic in this respect, as often so many different sectors have a direct stake in the overall activity. Perhaps the over-riding issue for marine ecotourism pertains to open access or, more specifically, to an assumption that marine resources may be sustained without any form of monetary or other cost. The reality that seas and oceans are, in effect, common property resources can lead to divergent notions of responsibility for and ownership of the marine environment.

Further to the open access issue, various legislative, interest and user conflicts can present themselves to marine ecotourism planners. These may be summarised as follows:

- overtly sectoral and uncoordinated legislation and policy;
- planning decision-making that is potentially isolated from marine ecotourism interests;
- incompatible decision-making with regard to infrastructure;
- rigid, bureaucratic planning systems that may have limited local adaptability;
- inadequate resource provision and statutory or informal local support from higher administrative levels;
- reluctance towards participation in voluntary mechanisms and initiatives (due to non-immediacy of direct benefits);
- potentially different national interpretations of international and national statutory measures (and hence different approaches to implementation);
- unpredictable effects of statutory and voluntary nature protection designations; and

- failure by land-use and transport planners to appreciate the dynamics of land–sea interrelationships and marine geographical boundaries.

Planning-related problems can also stem from management and research issues. Management of tourism in the marine environment may be based on a very limited understanding of marine scientific processes and wildlife characteristics and, furthermore, management approaches are potentially different between species and between management authorities. The limited resilience of marine and coastal species, habitats and landscapes to recover from serious mismanagement is a particular issue where tourism activities are part of the equation. In terms of research and monitoring, a lack of multidisciplinary approaches is observable, the effectiveness of regulatory and voluntary structures has not often been monitored, and research on marine species' tolerance and thresholds has typically been limited or lacking.

I will now discuss the particularly complex, 'messy' planning domain occupied by marine ecotourism activities and elaborate some of the issues identified earlier.

The nature of the marine environment

The dynamic and sensitive nature of the marine environment is perhaps the most challenging issue facing planners of marine ecotourism. Where terrestrial environments can often be spatially compartmentalised, for a number of reasons the same is not true of marine environments. The interconnectivity of the marine environment facilitates the transmission of harmful substances and their effects. Thus, marine and coastal areas are the ultimate recipients of the effects of environmental degradation and pollutants released into the air, land and water, including from streams and rivers. The sea then carries sediment, nutrients, pollutants and organisms through and beyond a specific location, all transported long distances by watercourses, ocean currents and atmospheric processes. Action taken in one marine locality, therefore, may affect another hundreds of kilometres distant and often nations apart (Cater & Cater, 2001).

Over-generous use of rivers and seas to disperse pollutant materials from urban development, industrial activity and the expiration of airborne pollutants can all impact on water quality (Taussik, 1997). In marine and coastal areas, this can substantially affect the use that can be made of the coastal and marine resources for marine ecotourism. The major threats to the health, productivity and biodiversity of the marine environment usually result from human activities on land – in both coastal areas and further inland. Most of the pollution load of the oceans, including municipal, industrial and agricultural wastes and run-off, as well as atmospheric deposition, emanates from such land-based activities and can have a negative effect on the marine environment. These areas are likewise threatened by physical alteration of the coastal environment, including the destruction of habitats of vital importance for ecosystem health. Ultimately, sustainable patterns of human activity in coastal areas depend upon a healthy marine environment, and *vice versa* (Intergovernmental Conference on Action for the Protection of the Marine Environment from Land-Based Activities, 1995).

The open nature of the marine environment also creates problems for planning regulation and policy. In particular, the uncontrollability and movement of marine wildlife within and above the water can limit the value of statutory planning structures. This condition exacerbates the existing problems with the enforcement of statutory and voluntary marine conservation measures. Despite designations such as MPAs (Marine Protected Areas – IUCN), marine nature reserves, e.g. Marine SACs (Special Areas of Conservation – EU Habitats Directive), or indeed of territorial waters, there are few physical barriers to accessibility. Given these circumstances, it is extremely difficult to police regulations in marine and coastal areas (especially in peripheral areas). The requirement for sustainability puts the onus on the provider to work with managers to ensure that impact on the ecosystem is reduced to a sustainable level, a process that is far from straight forward.

In terms of the open access issue, and responsibility for and / or ownership of the marine environment, Cater and Cater (2001) observe that the assumption that oceans and seas are common property resources is a problem. As they are not subject to individual or private ownership, there is a perception that they can be used without cost and problems arise, as this is an entirely correct perception in legal terms (for the most part). Under such circumstances, a particular problem also arises where there are positive incentives for individual or private users to exploit the resource to capacity or beyond, even if environmentally-damaging practices culminate in the degradation of the marine resource.

Further to ownership, there are various problems stemming from differing interpretations of international treaties / regulations and their designations by the individual state governments. Marine ecotourism is particularly susceptible to this issue, as by its very nature a wide range of regulatory structures must be acknowledged, from nature conservation to tourism, planning and others. Such problems are often exacerbated due to the nature of the authorities that have responsibility for implementing the regulations – often very different types of organisation within and between nations, all of which have different and possibly divergent objectives. Furthermore, when marine wildlife habitat areas span transnational boundaries, there may be problems with regard to different styles of planning systems between the countries concerned, making integration in planning even more difficult to achieve.

Spatial definition and jurisdictional coverage of the marine and coastal environment can present a challenge to planners. The coastal zone is the transitional area between land and sea, extending beyond the inter-tidal zone to include areas of land and sea where they influence one another and where use of the area impacts on the whole zone (European Commission, 1999). Unfortunately, this zone seldom corresponds to existing administrative or planning units and there are many spatial uncertainties in boundary and jurisdictional terms, having an impact on perceptions of local, regional and national responsibilities for the planning of marine ecotourism areas.

Use conflict and sectoral issues

On the notion of use conflict, the issue remains as to whether the protection of areas for nature conservation is as justifiable a use as is for example the development of a port and associated industrial activities or indeed the development of marine ecotourism. For example, attempts to balance nature conservation and visitor access can cause significant problems for those involved in marine ecotourism planning and management, though there is often a tendency to allow for more visitors, in the interest of increased income. This dilemma often presents itself as a trade-off. The historical relationship between tourism and protected areas has been heavy with doubts about the relative compatibility between tourism, nature conservation (Lawton, 2001) and, more strictly, nature preservation. However, the necessity for marine ecotourism to function as a symbiotic system, rather than a compromise, would mean that trade-offs between sectors are not the ideal nor would such practices be genuinely sustainable.

Ecotourism is particularly sensitive to the presence of other industries (Cohen, 2001) which, by implication, suggests that inter- and intra-sectoral integration is not easy to promote, let alone to achieve. As a consequence of inter-sectoral conflict in marine ecotourism (i.e. the issue of compatible *versus* incompatible industries), it is difficult to adopt an integrated approach to planning, especially when the impacts of the activity affect many different industries, authorities and interests. The lack of integration between sectors and interests is not unique to marine ecotourism, but due to its land–sea interface, maximum integration in marine ecotourism planning and management is usually rather difficult to achieve in practice. In addition, planning and management structures affecting terrestrial coastal areas may not be complementary to or may even conflict entirely with marine resource regulatory structures.

Various industries are incompatible with ecotourism but those that have been the most prominent in battles over the environment are mining, timber industries and ranching and – in the particular case of marine ecotourism – commercial fisheries, whaling, port industries and offshore oil extraction. The inter-sectoral conflict between nature conservation and resource extraction is particularly acute – as even if the extracted 'product' itself is harvested in an outwardly sustainable manner, the rest of the environment may be damaged. Furthermore, the process used to create a sustainable industry may also create visual and noise pollution: not only hostile to wildlife but also affecting the quality of the tourist experience (Cohen, 2001) and hence the willingness to pay, on the part of the tourist. This charge could be applied to some 'eco' tourism ventures, where trade-offs have been struck, to the detriment of the marine resource, the local community and/or the tourists. It is questionable whether such enterprises would constitute genuine ecotourism, where one or the other interests are compromised.

Cohen (2001) argues that zoning is a viable measure for tackling some of the problems caused by the intra- and inter-sectoral nature of ecotourism. He identifies two possible strategies for zoning ecotourism areas: first, where incompatible industries are prohibited from sharing a site; or second, where regulation restricts or expels activities currently operating in or near an ecotourism area. However, both of these are somewhat more difficult to consider in the marine context as opposed to the terrestrial context, particularly in terms of who can justifiably fight the battle of which incompatible industry should retreat. Returning to the issue of the openness of marine environments in pollution dispersal terms, plus the free mobility of marine wildlife, zoning is not always an appropriate measure, as marine ecotourism activities cannot be situated within discernable 'zones'. It is not, however, helpful to insist that marine ecotourism *must* operate in an area free of incompatible industries but certainly any *quid pro quo* alliance between marine ecotourism and the incompatible industry in question would not be a simple arrangement. Collaboration may not be entirely straightforward when the would-be collaborators have different training, loyalties, interpretations of the public interest and diagnoses of the resource problem (Parker, 2001) and this may even amount to an absolute divergence in the *raison d'être* of stakeholders at all levels. Aside from the relative difficulties involved, collaboration is, nonetheless, an important aspect of planning in the context of marine ecotourism.

This is not to assume, however, that the problems are purely inter-sectoral. Ecotourism planners and policy-makers must contend not only with incompatible sectors but also with other tourism activities that may interfere with ecotourism (Fennell *et al.*, 2001) and with few exceptions, it is the conventional tourism sector interests that prevail over ecotourism interests. Another potential intra-sectoral conflict even exists between the activities of marine wildlife tourism and marine ecotourism, where the legislative and policy provision for the former does not meet the requirements of the latter. This is problematic, when the basic free-ranging marine wildlife resource may consist of the same animal groups and habitats for both forms of tourism, even if operating from different terrestrial locations. An example of this problem lies in the southward movement of the Bottlenose dolphins in Scotland's Moray Firth in recent years – a population once thought to be resident there. Individuals are now being sighted as far south as St Andrew's Bay, on the east coast of Scotland, whereas in the first half of the 1990s, the dolphins were rarely spotted outside of the Firth (*BBC Wildlife Magazine*, August 2001). This means that the same group of dolphins could potentially be the core attraction for marine wildlife tourism and marine ecotourism enterprises alike, depending on where they are intercepted by the visitors. In this respect, marine wildlife tourism activities can serve to undermine the measures provided by genuine marine ecotourism ventures.

Sectoral responsibility and incompatibility are not the only boundaries to be considered by planners of marine ecotourism. As previously noted, when different national planning systems must function together in a marine environment that transcends national boundaries, the situation becomes highly complex and problematic. Remembering that marine wildlife does not necessarily remain within a given marine geographical location (and nor does marine pollution), transnational planning considerations are very important. However, in practice, different national and regional planning systems operate in differing ways and with varying objectives in their regulation. In many places in the EU Atlantic periphery, for example, the need for economic development is paramount. The planning processes here may promote development but may not seek to minimise its negative environmental impacts in all of the regions that are coveted by the highly mobile marine wildlife species. Furthermore, some regions have made more progress on embracing sustainable development objectives than others and while there may be provision for plans at various levels, they may not be complete or up to date, as the relative importance and rigidity of planning instruments can vary a great deal from country to country.

Often, the authorities that have responsibility for land-use planning and coastal/marine planning are not necessarily based within the same organisation. Other government bodies, agencies and even non-governmental organisations (NGOs) are often also involved in the equation. In West Clare (Ireland), the resident population of Bottlenose dolphins in the Shannon estuary is potentially affected by legislation and policy not just from government departments but from agencies such as the Dúchas (the Irish Parks and Wildlife Service), An Taisce (the National Trust for Ireland), Bord Failte Eireann (the Irish Tourist Board), the national Environmental Protection Agency and the national Marine Institute. Regional bodies such as Shannon Development and NGOs such as the Shannon Dolphin and Wildlife Foundation, Birdwatch Ireland and the Irish Landmark Trust also feature in the equation. The challenge often lies in integrating the efforts and interests of the many different organisations usually involved (or with a stake) in marine ecotourism planning.

In some parts of the world, the ownership and/or stake of indigenous peoples in marine resource management can complicate even further the planning process for activities such as marine ecotourism (see Orams, this volume). However, in some cases, this has been well managed and is mutually beneficial to the stakeholders and the environment concerned. In the Great Barrier Reef Marine Park and UNESCO World Heritage Site, Australia, there is a significant indigenous cultural relationship with the Reef that long pre-dates non-indigenous involvement in the area. Contemporary indigenous peoples are attempting to retain their cultural association with, values and use of the area in the face of increasing pressure from coastal development, commercial fishing, private recreational use and rapidly increasing tourism use. With this in mind, the Great Barrier Reef Marine Park Authority (GBRMPA) established an Indigenous Cultural Liaison Unit (ICLU) in 1995. The aim was to address the interests and needs of indigenous peoples in relation to governance, and to maintain the cultural and traditional values associated with the Great Barrier Reef. The Unit addresses issues such as the recognition of cultural heritage values, semi-subsistence resource use, information-sharing, cooperative management, protocols, cultural advice and liaison (GBRMPA, 2002).

On the whole, planning policies do not always respond in time and effectively to the development process of marine ecotourism and rarely have the over-riding aim of maintaining functions and using resources in a sustainable manner. As with ICZM, marine ecotourism stakeholders do not always view concern for socio-economic values and the preservation of environmental quality as integral parts of a single system. Local planning policy may be unable to reflect the complexity of the planning process, particularly when it should consider a variety of stakeholders representing different views. In addition, those involved in planning for recreation and tourism have often treated marine ecotourism activities in isolation from other factors that make up the social, environmental and economic fabric of a region. A lack of recognition of such factors is particularly acute in terms of peripherality. In core regions, it is often the case that environmental concerns are allowed to prevail and be considered realistically but in the more peripheral regions, in the name of narrowing the gap between core and periphery, the social and economic concerns often dominate over the environmental, having an effect on the likelihood of genuinely sustainable marine ecotourism. As it is probably the case in many countries that the best sites for marine ecotourism development are located in the more peripheral areas, a major challenge is presented to planners in reconciling the objectives of peripheral communities with the presence of marine ecotourism.

Regulation issues

Conflict may also arise when the policy priorities of the various national government agencies with a stake in marine ecotourism are not complementary. The primary missions of the agencies involved may deal with different functions and may have stronger loyalties elsewhere (Parker, 2001). This can be the case with departments that can impact on marine ecotourism (such as public works, transport, environment, cultural affairs, rural affairs, investment and economic development and education, as well as planning). More traditional governmental sectors may serve as advocates of primary industries and (particularly in less developed countries) may enjoy more political power (Hall, 1994). This is opposed to the circumstances of government agencies serving tourism and environmental protection, for example, which tend to be politically weak and less effective in advancing their agendas. Marine ecotourism planning interests are therefore left in a potentially precarious policy situation. This is not only because the government agencies it relies upon have such relatively weak constituencies but also because these agencies are less likely to gain local support than other sectors. As such, the tourism sector is placed in the disadvantageous position of having to react and adapt to, rather than to influence, the external policy forces within its domain. In addition, within national government, tourism and other policies often have overlapping life expectancies. When the overall policy changes following a national election, any national-scale policy for tourism, or indeed ecotourism, may only be halfway through a five-year plan (Fennell *et al.*, 2001).

A number of negative consequences can flow from public commitment to infrastructure development; for example, many of the negative impacts (including the growth of nature-based tourism) on Australia's Great Barrier Reef can be traced back to a single infrastructure decision – the construction of Cairns International Airport – in 1984 (Parker, 2001). Indisputably, however, these public works are an essential part of the planning and policy context within which marine ecotourism operates. In infrastructure terms, governments often tend to be considerably less sympathetic to ecotourism than to the traditional variety of tourism, due to considerations relating to economies of scale – concentrating resources in large resort areas appears to them to produce a much more efficient use of capital resources (Inskeep, 1987). Considerations of scale predispose decision-makers in the direction of unsustainable conventional tourism and often, 'bigger is better' to investors and planners, because this may enable distribution of fixed costs over a larger number of units and visitors.

In some parts of the world, there have been imbalances in regulating for species protection and other marine and tourism-related policy areas. Regulatory structures are often specifically geared to the protection of marine species and habitats but the regulations do not generally address other issues relating to ecotourism, notably the promotion of tourism, environmental management, people safety and the commercial viability of tourism operations. As such, resource managers often have conflicting responsibilities. Statutory authorities may have little or no requirement to balance commercial development with the protection of marine species and habitats: nor are they tied up in issues that are perceived as peripheral to their main concern: marine wildlife protection.

Nature protection designations can also have unpredictable after-effects which may serve to undermine their usefulness altogether. Although regulation through conservation designations may be important for protecting the marine environment, it may also have substantial socio-cultural impacts about which very little is currently known or acknowledged. The potential flip side of marine nature protection legislation is that a new designation can stimulate vast interest in a previously low-profile habitat or species, with the consequence of escalating visitor numbers. Therefore, designation of a marine wildlife site may end up being the catalyst for increasing visitor flows and hence increasing pressure on such sites.

Badalamenti *et al.* (2000) observed a general increase in tourist activities in Marine Protected Areas (MPAs) in the Mediterranean following their formal designation as such. A large increase in the number of divers and vessels using MPAs has already had impacts on natural benthic communities as a result of diver damage, mooring and the feeding of large fish by divers in the area.

Only in a few MPAs has any emphasis been given to promoting public awareness of negative impacts. Although nature conservation should be considered the fundamental objective of MPAs, neglecting their social, cultural and economic impacts has, at times, led to problems in achieving local consensus, if not outright hostility, on the part of the local community towards the development of marine ecotourism.

Research and monitoring issues

There has been a general dearth of inter- and multi-disciplinary approaches to research and monitoring, which could be used to inform the planning process about the reality of marine ecotourism activities. Planning and managing for ecotourism in the marine environment should ideally be conducted on a multidisciplinary basis (Badalamenti *et al.*, 2000) but a problem arises when no single model can be considered universally valid, for example for all MPAs worldwide. The highly variable characteristics of marine and coastal areas requires that different weightings be assigned for each factor in order to achieve genuinely sustainable and symbiotic marine ecotourism, moving towards a lasting consensus, in favour of conservation and appropriate local benefits.

Another challenge for planning decision-makers concerns cooperation with marine ecotourism site managers in establishing thresholds and levels of species and habitat tolerance to adverse impacts from tourism. This is particularly difficult, as many of the species on which marine ecotourism is based are relatively under-researched, in terms of their sensitivity and tolerance to tourism and other development pressures. Marine wildlife is extraordinarily difficult to study compared with terrestrial wildlife and such work is typically costly of time and resources. Consequently, there has been a general lack of knowledge of the target species' biology and behaviour, especially in terms of the direct effects of tourism. Information regarding the location and population dynamics of target species is often lacking, and this extends to a lack of knowledge regarding the impacts that the development of ecotourism may have on such populations (META-Project, 2000).

Regulating for the protection of specific species can have advantages but, although theoretically ideal, its usefulness may be limited in practice. This is because the multiple individual regulations become increasingly confusing and unwieldy. Furthermore, species-specific regulations are problematic when most touristic viewers of marine wildlife, and even marine tourism operators, cannot always positively identify one species from another and are hence uncertain of how to behave. This can extend to voluntary structures, when generic codes of conduct are not able to give species-specific advice to visitors.

Planning processes for marine ecotourism ideally require continual monitoring, from the effectiveness of overall structures to the behaviour of the animals targeted and the degree of compliance with the regulations and voluntary codes by operators. However, the systematic monitoring and review of such regulatory and voluntary structures can be expensive and cumbersome, particularly when the conditions under which structures are devised are often dynamic and difficult to conceptualise for practical purposes.

Having looked at the problematic issues in planning for marine ecotourism, the next section will examine some areas in which such problems can be addressed, reducing the negative effects to a manageable level.

Addressing Planning Problems in Marine Ecotourism

So far, this chapter has set out a variety of problems, with few solutions, and it could be argued that, realistically, solutions are thin on the ground at present. However, there are various ways in which several of these problems can be addressed, if not entirely then at least in terms of ameliorating the negative effects. Measures for overcoming the barriers in formal planning systems do exist and have been implemented successfully in various parts of the world, although successes and other experiences have not often been tied down to tangible, transferable frameworks and guidelines. Fortunately, this appears to be changing.

It is clearly the case that existing statutory and other formal mechanisms have an important role to play in planning and managing for genuinely sustainable marine ecotourism. However, it is important to recognise that these will often need to be supplemented by a range of informal and voluntary measures in order to achieve a balance between top-down and bottom-up approaches. Self-regulation is an important part of planning for marine ecotourism. In some cases, the voluntary structures have evolved into statutory provisions due to high levels of support, commitment and lobbying. For example, an absence of government legislation in Antarctic marine tourism led to cooperative self-regulation among tour operators and principles for management were developed and adhered to. Cruise operators later formed a coalition, the International Association of Antarctica Tour Operators (IAATO), which issued guidelines and codes of conduct. Later, the Antarctic Treaty issued formal recommendations for visitors and similar guidelines of its own but it was the industry and not the regulating body that first set the precedent for regulation. More recently, some operators have moved beyond the IAATO guidelines and produced their own manuals of good practice, conforming to Treaty regulations and, where possible, anticipating future legislation (Stonehouse, 2001: 231–2).

It is possible, therefore, for a lack of (or limitations of existing) regulations to be mitigated by voluntarism, particularly where the economic benefits associated with tourism can be aligned with conservation. Supplementing the statutory approach with voluntary regulations can address the shortcomings that formal regulatory structures tend to exhibit, in respect of the unique problems inherent in planning for marine ecotourism.

The educational aspect is also an important consideration (see the chapters by Townsend and by McDonald & Wearing in this volume). Ecotourism confers a special responsibility on ecotourism experience providers to ensure that they include suitable and effective interpretation of the marine environment in which the ecotourism activity is taking place. Designed well, educational devices can help to ensure that voluntary and statutory planning measures are respected and more closely adhered to.

Collaboration is also an important means of building planning solutions for marine ecotourism. Development, planning and management processes in genuine marine ecotourism all occupy a complicated problem domain that requires collaboration, rather than simply cooperation. Problems associated with marine ecotourism can be so complex that it is beyond the capability of any single individual or organisation to resolve them. Collaboration between stakeholders in the problem domain is therefore essential, if marine ecotourism is to be developed to be genuinely sustainable. In the Great Barrier Reef Marine Park and UNESCO World Heritage Site, Australia, planning occurs through a system of integrated management administered by the GBRMPA. Without collaboration between the numerous bodies involved, monitoring, planning and day-to-day management of such a vast and diverse Marine Park would be impossible. Planning involves a wide range of councils, agencies and committees and although GBRMPA is the principal adviser to the Commonwealth Government, many other government agencies and NGOs actively participate in the management of the Marine Park (GBRMPA, 2002).

Many coastal planning authorities work to models based on the assumption that development is a zero sum game, in which developers and government are locked in combat to win the development game (Potter, 1996). It is now generally conceded that developers, governments and both private and public sectors must win, in order for the people to benefit from new investments. An activity such as marine ecotourism, which depends on high-quality natural resources can benefit considerably from win–win negotiating systems but unfortunately win–win concepts have not yet been widely applied to the actual processes of planning, coordinating and permitting ecotourism development (Potter, 1996). Experience from the Soufriere area of St Lucia, for example, indicates that win–win planning processes can be effective. In this area, major, long-term tourism development planning processes have successfully involved local residents, fishermen, hotel owners and planners and a variety of government agencies. The effort has been heavily supported by outside interests, including the USAID / ENCORE programme implemented by the Natural Resources Management Unit of the Organisation of Eastern Caribbean States (Potter, 1996).

Marketing is another area in which planning measures can be complemented, with careful design and implementation. Marketing that does not prioritise environmental protection can compromise even the most well-planned and meticulously managed marine ecotourism activity. This is because the marketing of marine ecotourism experiences may be in contradiction with the associated planning and management objectives, particularly in that it is likely to result in excessive visitor numbers or inappropriate behaviour on the part of the tourists. The same is also true of community participation in the marketing process, for unless the community is sufficiently in control of the marketing of the activity, marketing efforts are likely to run counter to the planning and management of the activities being promoted (see Hoctor, this volume). Responsible marketing in this context should embrace the notion of environmental and socio-cultural stewardship. In any case, the marketing of marine ecotourism should be just as consistent with the principles of sustainability as the planning and management efforts.

Continual monitoring can assist in assessment of whether or not marine ecotourism is being effectively planned and managed: not only in terms of the behaviour of the animals targeted but also the degree of compliance with regulations by the commercial operators. It may also be necessary to review systematically any regulatory or voluntary structures through a monitoring process. The operating conditions brought in by regulation may be successful in protecting the marine ecotourism resource base in terms of the short-term, day-to-day effects of marine ecotourism. However, the long-term effects of marine ecotourism also need to be subject to monitoring and periodic review. The assessment of possible longer-term, cumulative impacts remains a challenge for all marine ecotourism stakeholders, both in terms of the resource and of the benefits to communities.

With a few exceptions (such as the Antipodes and North America), national ecotourism policy is seldom explicit, particularly in Europe (Fennell *et al.*, 2001). There is a need for measures that are, to a degree, prescriptive, but that are always be interpreted in the local context and with the local stakeholders, without whose active support and participation, no marine ecotourism would be genuinely sustainable.

The most effective planning instruments at all levels tend to incorporate appropriate policies that support marine ecotourism, while ensuring that planning officials (and elected members who prepare plans and issue permits) fully understand the nature and value of marine ecotourism. Success may depend on early contact between initiators of marine ecotourism ventures, the planning authority and other stakeholders when any new project comes forward. In the ideal scenario, local planning policy-makers and initiators of local marine ecotourism projects would review the planning policy context and, in all cases, it is important to ensure that any national and regional planning instruments are supportive of appropriate marine ecotourism.

Despite the problems discussed here with zoning as a planning measure, some areas of large-scale marine ecotourism have managed to implement zoning systems effectively. In the Great Barrier Reef Marine Park, zoning allows areas that need permanent conservation to be protected from potentially threatening processes; by being placed 'off limits' to users (except for the purpose of scientific research) for varying lengths of time. Specific management plans are prepared for intensively used or particularly vulnerable groups of islands and reefs, and for protection of vulnerable species or ecological communities. These plans complement zoning by addressing issues specific to an area, species or community on the reef, in greater detail than can be accomplished in the broader, reef-wide zoning plans (GBRMPA, 2002).

Voluntary and public-sector ecotourism organisations can also play a major role in reducing the burden on planning systems in making ecotourism a more sustainable tool for conservation and community development (Halpenny, 2001, and this volume). Ecotourism organisations can be administrative or functional structures that are concerned with ecotourism and their roles can range from grassroots advocacy to international policy-making, initiated by membership NGOs, non-membership NGOs and public-sector/government agencies. In terms of policy-making from ecotourism partnerships, however, most success has been observed at the micro scale (Parker, 2001), with a focus on issues such as financial incentives, transportation decisions and very specific environmental management practices.

The scale of marine ecotourism can also determine the degree of difficulty in planning, particularly with respect to mitigation of negative impacts. From the operational perspective, research suggests that those ecotourism providers that are most effectively practising marine ecotourism tend to: operate in relatively remote areas; have some evolving environmental and tourism management structures in place; and be run by self-motivated operators (Harris & Leiper, 1995). However, as an ecosystem-based form of tourism, marine ecotourism also has an obligation to respect the global environment, which may be adversely affected by the global warming implications of the transport needed to reach remote areas. Broadly speaking, local planning policies would be better informed it they could establish an understanding of both the likely success criteria and the conditions for genuine sustainability in marine ecotourism, in its various environmental, economic, social and cultural aspects.

In any eventuality of striking a local balance in formal and informal mechanisms, a clear policy statement at the national and/or regional levels would give an important signal back to local decision-makers. At the local level, there may be opportunities to incorporate revised policies and criteria for marine ecotourism during the periodic review of plans. In these cases, it may be helpful to develop some community-based draft criteria, or 'model policies', which the planning authority could consider for formal adoption. Where this is not possible, it may be appropriate to prepare community-based supplementary guidance for decision-makers, which provides additional advice and criteria tailored to local circumstances. It is particularly important that the implications of any binding planning instruments are considered, since their revision may be a lengthy and complicated process.

Conclusion

Planning, as a process, is undeniably central to marine ecotourism activities. However, it is clear that there are several issues bound up in planning processes, in the context of marine ecotourism. These result not only from the particular nature of the marine environment as opposed to the terrestrial but also from specific use conflicts and sectoral issues, regulation issues and research and monitoring issues inherent in marine ecotourism.

On the whole, the challenge lies in addressing these problems and achieving genuinely sustainable marine ecotourism within existing planning structures, while seeking to initiate voluntary control mechanisms where statutory structures fall short of requirements. In any case, it is fundamental that locally-generated, community-based policies for the planning and management of marine wildlife, tourism, resources and coastal zone management go beyond the statutory and equivalent systems. This is, however, not to deny the importance of pressing for change in national and international regulation and policy related to marine ecotourism, or at least for recognition of marine ecotourism as a phenomenon to be planned for, at all levels.

While there may be observable weaknesses in current marine ecotourism-relevant planning policy and practice, it may not be realistic to change the statutory mechanisms entirely. Nor, however, would it be realistic to avoid addressing the inherent weaknesses in such mechanisms. In the interim, given the difficulties in initiating widespread and standardised change to planning systems, a careful blend of statutory and voluntary structures could alleviate the messy and complex nature of the marine ecotourism planning problem domain.

This chapter has set out various problems and difficulties faced by planners, not least that marine wildlife, like marine negative impact, is highly mobile and not contained within any given boundary. Future work should be directed towards solutions for the problems of existing international, national and regional legislation, while permitting planners to broadly work within it through supplementary voluntary measures. This will be increasingly appropriate, should marine ecotourism continue to grow internationally.

Acknowledgements

Some of the material presented in this chapter was originally prepared for the EU Interreg IIc Project Marine Ecotourism for the Atlantic Area META- (Project EA-C1UK-No.3.14). The author would like to thank Brian Garrod, Vincent Nadin and David Bruce for their support in the preparation of this chapter.

References

Badalamenti, F., Ramos, A. and Voultsiadou, E. (2000) Cultural and socio-economic impacts of Mediterranean Marine Protected Areas. *Environmental Conservation* 27 (2), 110–25.

BBC Wildlife Magazine (2001) Seeing fins. *BBC Wildlife Magazine* 19 (8), 62–4.

Cater, E. and Cater, C. (2001) Marine environments. In D. Weaver (ed.) *The Encyclopaedia of Ecotourism*. Wallingford: CABI.

Cohen, J. (2001) Ecotourism in the inter-sectoral context. In D. Weaver (ed.) *The Encyclopaedia of Ecotourism*. Wallingford: CABI.

European Commission (1999) *Towards a European Integrated Coastal Zone Management (ICZM) Strategy: General Principles and Policy Options*. Luxembourg: Office for Official Publications of the European Communities.

Fennell, D.A., Buckley, R. and Weaver, D.B. (2001) Policy and planning. In D. Weaver (ed.) *The Encyclopaedia of Ecotourism*. Wallingford: CABI.

Garrod, B. (2002) Monetary valuation as a tool for planning and managing ecotourism. *International Journal of Sustainable Development* 5 (3), 353–71.

GBRMPA (2002) Management of the Great Barrier Reef Marine Park / World Heritage Site, Australia, (online). [http://www.gbrmpa.gov.au/] (Accessed February 2002).

Hall, C.M. (1994) *Tourism and Politics: Policy, Power and Place*. Chichester: Wiley

Halpenny, E. (2001) Ecotourism-related organisations. In D. Weaver (ed.) *The Encyclopedia of Ecotourism*. Wallingford: CABI.

Harris, R., and Leiper, N. (1995) *Sustainable Tourism: An Australian Perspective*. Chatswood: Butterworth-Heinemann.

Inskeep, E. (1987) Environmental planning for tourism. *Annals of Tourism Research* 14 (1), 118–35.

Intergovernmental Conference on Action for the Protection of the Marine Environment from Land-Based Activities (1995) *Global Programme of Action for the Protection of the Marine Environment from Land-Based Activities*. (Adopted on 3 November 1995.) Washington DC, USA.

IUCN (1992) *Protected Areas of the World: A Review of National Systems: Neoarctic and Neotropical* (Vol.IV). Gland: World Conservation Union.

Lawton, L. (2001) Public protected areas. In D. Weaver (ed.) *The Encyclopaedia of Ecotourism*. Wallingford: CABI.

META- Project (2000) *Marine Ecotourism for the Atlantic Area (META-): Baseline Report*. Bristol: University of the West of England.
Parker, S. (2001) The place of ecotourism in public policy and planning. In D. Weaver (ed.) *The Encyclopaedia of Ecotourism.* Wallingford: CABI.
Potter, B. (1996) Win–win coastal planning. In B. Potter / Island Resources Foundation (eds) *Tourism and Coastal Resources Degradation in the Wider Caribbean.* Washington, DC: Island Resources Foundation.
Stonehouse, B. (2001) Polar environments. In D. Weaver (ed.) *The Encyclopaedia of Ecotourism.* Wallingford: CABI.
Taussik, J. (1997) The influence of institutional systems on planning the coastal zone: Experience from England / Wales and Sweden. *Planning Practice and Research* 12 (1), 9–20.

Chapter 4

An Assessment of the Framework, Legislation and Monitoring Required to Develop Genuinely Sustainable Whalewatching

SIMON D. BERROW

Introduction

Whalewatching, defined by the International Whaling Commission (IWC) as any commercial enterprise which provides for the public to see cetaceans (whales, dolphins and porpoises) in their natural habitat (IWC, 1994), is one of fastest growing tourism products in the world. Already estimated in 1998 to be worth €1.12 billion worldwide, the industry is still growing at 12.1% per annum (Hoyt, 2001). Whalewatching in Europe is relatively new but expanding rapidly. Average rates of 8.8% growth per annum between 1991 and 1994 have increased to 19.6% between 1994 and 1998 (see Table 4.1).

The data shown in Table 4.1 refer to expenditures for whalewatch tickets (direct expenditures) and associated expenses incurred by tourists during, as well as immediately before or after, whalewatching (indirect expenditures). A conservative estimate of the total expenditures from whalewatching near urban centres with day (or less) trips is 3.5 times the direct expenditures (Kelly, 1983; Hoyt, 2001). In remote centres, which require more spending on travel, food and accommodation, total expenditures are usually at least 7.67 times the direct expenditures (Duffus, 1988; Hoyt, 2001). For the most part, the 3.5 and 7.67 factors stand up to inflation, as the ticket prices increase at approximately the same rate as the other expenses.

Table 4.1 Increase in the number of whalewatchers and their expenditure (€) in Europe (from Hoyt, 2001)

Year	*No. of whalewatchers*	*Direct expenditures*	*Total expenditures*
1991	158,763	2,532,000	6,373,000
1994	204,627	4,618,000	24,623,000
1998	418,332	12,373,000	51,552,000

Whalewatching is becoming economically important in many EU countries. However, the extent to which tourism activities can have a detrimental effect on the behaviour of whales and the long-term sustainability of whalewatching has not been assessed. If whalewatching, along with other forms of marine ecotourism, are to be viable economic alternatives for coastal communities in Europe, it is essential that such activities are sustainable.

The scientific management of whalewatching and tourism is extremely limited. There are few published studies of whalewatching operations (e.g. Au & Green, 2000; Berrow & Holmes, 1999; Corkeron, 1995; Findlay, 1997; Leaper *et al.*, 1997) and, despite the economic importance of whalewatching, there have been few socio-economic studies of this industry. Although there are many other associated benefits, such as conservation and education, whalewatching is an economic activity whose principal objective is financial and requires a return on investment. To be genuinely sustainable, whalewatching should be both economically and ecologically sustainable.

Potential Frameworks for Sustainable Management

Regulations and legislation

A variety of voluntary and legislative measures have been used to manage whalewatching throughout the world. All countries in Europe have national wildlife legislation, which tends to address issues of harassment, disturbance and direct killing. However, amendments to this legislation are necessary in most countries to manage tourism activities such as whalewatching. In 1994, the International Whaling Commission (IWC) established a Whalewatching Working Group and considered that there was a general view on the need for regulations to provide adequate safeguards for the whales, as voluntary guidelines or codes of conduct may not always be strong enough as controls (IWC, 1995).

The conservation management of cetaceans is covered by a number of international conventions ratified in Europe, including the Berne and Bonn Conventions and CITES. The OSPAR Convention may also have potential for resource management. Within Europe, the Habitats Directive is increasingly being identified as most relevant, as it provides for designation of Marine Protected Areas (MPAs) and thus, the regulation of whalewatching within these designated sites. However, outside of (MPAs), additional legislation may have to be considered in order to provide a legal framework specifically for tourism management. The most successful management potential exists where the number of operators and vessels are licensed.

Marine Protected Areas

MPAs can play a strategic role in the management of marine environments and may be designated for a variety of reasons (see Kelleher & Kenchington, 1992). They are increasingly being considered as a framework for managing whalewatching. For example, Stellwagen Bank, off Northeast USA, is one of the most important whalewatching sites in the world, with at least 10 million whalewatchers between 1975 and 1993. The Stellwagen Bank National Marine Sanctuary (http://www.sbnms.nos.org) was established in 1993, in recognition of its importance for whales, largely determined from work carried out on whalewatching vessels, and the threat of excessive disturbance from the whalewatching industry.

'No go zones' are often a feature of MPAs and can aid management by providing areas for whales to be free from disturbance by whalewatching. Critical habitats such as calving areas and rubbing beaches (for orcas, *Orcinus orca*) are often those areas designated as 'no go zones'.

In Europe, the Habitats Directive requires member states to designate sites for the conservation of specific species and their habitats, under the Natura 2000 network. All cetacean species are listed under Annex IV and two species, the bottlenose dolphin (*Tursiops truncatus*) and the harbour porpoise (*Phocoena phocoena*), are listed under Annex II – species whose conservation requires the designation of Special Area of Conservation (SAC). Bottlenose dolphins are frequently the 'target' of whalewatching, as they are coastal and approach boats readily. Whalewatching in Europe on this species occurs in the Azores, Canary Islands, Croatia, France, Greece, Ireland, Italy, Portugal and the UK.

MPAs have also been established in Europe outside the framework of the Habitats Directive. The Ligurian Sea Sanctuary was established by France, Italy and Monaco in 1999. Although mainly concerned with industrial and fisheries impacts, this MPA will also attempt to regulate whalewatching operations.

MPAs require adaptive management; that is, a form of management based on solid scientific grounds and on performance studies, allowing a systematic evaluation to be made of the degree to which management objectives have been attained. In such a context, scientific research plays a strategic role.

Codes of conduct/guidelines

Models of best practice, including codes of conduct and accreditation schemes, are increasingly being promoted for the management of whalewatching. Most countries and communities involved in whalewatching have some regulations, including codes of conduct with which whalewatching operators are asked to comply. Often, these are voluntary, but have legal enforcement in some areas, which may be through local bylaws or within the wider legislation of MPAs.

Carlson (2000) reviewed international whalewatching guidelines, including six countries in Europe (Azores, Canaries, France, Ireland, Norway, UK). Carlson's review suggests that a wide range of guidelines have been introduced, including restrictions on the number of boats close to cetaceans, a minimum approach distance and sometimes a maximum time allowance during each encounter. However, the IWC Whalewatching Working Group suggests voluntary guidelines or codes of conduct may not always be strong enough to control whalewatching activities, especially if there are management conflicts (IWC, 1995).

Both Portugal (Azores) and Spain (Canaries) have recently approved new regulations especially for the management of whalewatching. In the Azores and Canaries, companies dedicated to whalewatching must apply for a permit and in the Canary Islands all vessels must carry a monitor-guide to ensure codes of conduct are respected. In the UK and Ireland, guidelines for whalewatching operators have been published but no specific regulations have been passed. In an MPA such as the Shannon estuary, Ireland (a SAC for bottlenose dolphins), whalewatching is a notifiable activity and operators, as well as recreational craft, are obliged to adhere to the code of conduct (see Berrow, this volume).

Voluntary guidelines, often developed by operators, can be more restrictive and comprehensive than those included in a legislative framework. For example, the boater guidelines promoted by the Northwest Whalewatching Operators Association in the Pacific Northwest – a favoured area for orcas – (http://www.nwwhalewatchers.org/whalewatchguidelines.htm) are promoted beyond the boundaries of the San Juan Islands National Wildlife Refuge and Wilderness Area.

Essential for the scientific management of whalewatching is hypothesis testing. International Fund for Animal Welfare (IFAW) *et al.*, (1995) have produced a framework for assessing the effectiveness of each parameter used to determine impact (Figure 4.1). However, the IFAW (1996: 29) warns:

> it is very difficult to provide more specific recommendations, a priori, as to what would constitute biologically significant impacts warranting changes to the regulations or guidelines. Particular care needs to be taken to design studies so that they may have the best possible chance of detecting any changes in behaviour. However, the participants emphasised that a failure to detect changes in behaviour would not necessarily mean that such changes were absent.

Research

Due to the lack of basic information on the ecology of cetaceans and the impact of tourism, research should be an essential element in the sustainable management of whalewatching. Research should not be seen as having a negative impact on whalewatching: indeed Tilt (1985) found that in California, whalewatchers were willing to pay more if some of the tour proceeds went towards whale research or education.

Developing sustainable whalewatching requires an inter-disciplinary approach between the biological and social sciences, to formulate management plans that

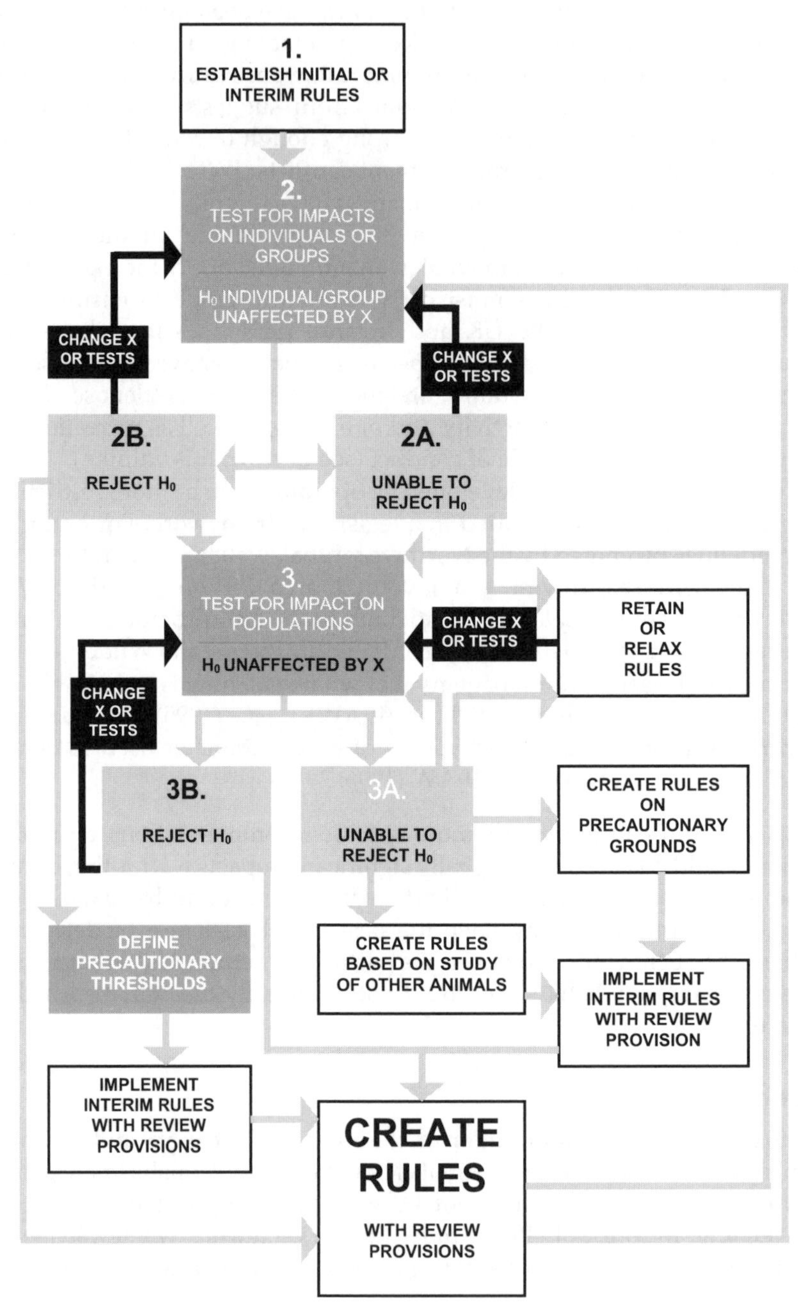

Figure 4.1 Making decisions about whalewatching rules
(from IFAW *et al.*, 1995)

promote wildlife encounters for tourists without harming wildlife. Research needs to establish

> behavioural and reproduction benchmarks that will allow managers to recognise when the focal species is being disturbed, and if that disturbance has potential to harm the individual or the population. Traditional wildlife management agencies may not be equipped to incorporate social science research. (Duffus & Dearden, 1990: 227)

Ideally, baseline research on the distribution and relative abundance of cetaceans should be carried out prior to the development of whalewatching (e.g. Berrow *et al.*, 1996). In reality, this has rarely been possible, and often it is the availability of whalewatching vessels that facilitates research projects, which would not be generally possible without this facility.

Carrying capacity

The carrying capacity of whales to whalewatching is the ultimate constraint to sustainable whalewatching and all activities should be carried out within it. Unfortunately, there is little or no information on the carrying capacity of whales to whalewatching. IFAW *et al.* (1995) list some biological and population parameters that may be impacted by whalewatching. To assess carrying capacity, the most sensitive parameter must be determined, its limits assessed and whalewatching managed to within these constraints.

If carrying capacity exceeds demand, then whalewatching could be sustainable. If, however, demand exceeds carrying capacity, then management intervention will be necessary.

For example, if the amount of time vessels are within 100m of cetaceans is the most sensitive variable, then whalewatch operations must work within these limits. This does not necessarily mean that the industry is overly constricted but it does mean that the industry must restrict this element of operation to within sustainable limits. This could mean limiting the amount of time vessels are within 100m on each trip.

Education

Similar to ongoing research, education should be an integral part of developing sustainable whalewatching. Information on the species and habitats being exploited should be available to whalewatchers and operators alike. Information on the legislation and codes of conduct etc. should be promoted at all opportunities, together with the sensitivity and conservation value of the site.

Stakeholder involvement

Increasingly, the involvement of stakeholders in resource management is being viewed as critical to the success of sustainable development (Scheyvens, 1999). Stakeholders, including state bodies, local authorities, local community groups and private companies, should be identified and invited to contribute to the process.

Responsibilities and aspirations should be agreed upon and a development plan should be formulated, including timescales and the resources required for implementation.

At sites where there are significant numbers of whalewatch operators, some have formed associations or organisations. For example, there are 36 operators from Washington State and British Colombia who are members of the Whalewatching Operators Association NW (http://www.nwwhalewatchers.org). Such associations and organisations may be helpful in facilitating meaningful participation on the part of whalewatch operators.

Funding

Sustainable development of whalewatching requires a long-term funding commitment. State agencies are unlikely to, and should not necessarily be expected to, fund the monitoring of whalewatching activities in the long-term. The 'polluter pays principle' has been applied to industrial development throughout Europe and may be appropriate for the whalewatching industry.

The whalewatching industry could generate the funds necessary for monitoring their activities through membership fees of operators associations or a trip levy. Some organisations, such as the Whalewatching Operators Association NW, fund research, which provides them with the information they can use to educate passengers and the public and themselves. Swim-with-dolphin operators in the Bay of Isles in New Zealand also pay a levy, which is used to fund research and monitoring. The funds are administered by a Committee comprising operators, scientists and staff from the New Zealand Department of Conservation, and although this levy is not legally binding, the operators contribute freely, as they can decide what research is funded.

A number of studies have shown that willingness to pay (WTP), when based on levies, is much higher if the funds are demonstrably used for research and monitoring (Tilt, 1985; Orams, 2000). Research, education and monitoring should not be seen as luxuries but as necessary parts of the sustainable management of whalewatching. Such activities should be considered part of the operating overheads involved in operating whalewatch tours, similar to boat fuel and insurance. Unless funding for research, education and monitoring is built into the operating costs of the whalewatching industry, it is unlikely that whalewatching can become genuinely sustainable.

Monitoring Indices to Assess the Sustainability of Whalewatching

Despite the economic importance and longevity of whalewatching in many parts of the world, there is surprisingly no long-term monitoring of the whalewatching industry and its effects on cetaceans at any whalewatching location. There have been a number of short-term studies to assess the effect of tour boats and other activities (e.g. swim-with-dolphin operations, (Samuels *et al.*, 2000)) on cetacean behaviour but no ongoing monitoring. There are data being collected as part of other studies, which could be used to address tourism-related issues but they are not designed to assess impact.

Biological monitoring

In order to develop genuinely sustainable whalewatching, the effect of tourism activity on the species and habitat being exploited must be quantified and the impact assessed. This information is essential to determining carrying capacity, which is the amount of activity a species or habitat can be subjected to without affecting its long-term viability, and is the biological framework within which whalewatching is constrained. In practice, it is extremely difficult to quantify carrying capacity and this has not been achieved at any whalewatching location in the world, although some locations are attempting to address this issue (e.g. Shark Bay, Australia). The studies referred to here are not meant to be exhaustive but indicative of the type of work carried out and the results available.

Short-term studies

The reactions of whales to whalewatching may be negative, neutral or occasionally positive. One of the first studies to assess the reaction of whales to whalewatching vessels was carried out by Gordon *et al.* (1992) who studied sperm whales (*Physeter macrocephalus*) in Kaikoura, New Zealand. Indeed, there has probably been more work carried out on the effects of whalewatching on sperm whales than any other species (IFAW, 1996). This work showed that although there was some effect of whalewatching vessels, it was considered minimal and the licensing authority, the New Zealand Department of Conservation, issued additional whalewatching licenses based on this advice. Blane and Jaakson (1994) recorded avoidance responses by belugas (*Delphinapterus leucas*) to tour boats in the St Lawrence River, Quebec, Canada. These included bunching together, longer dives and shorter surfacing time. Janik and Thompson (1996) reported similar avoidance reactions for bottlenose dolphins in the Moray Firth, Scotland. Indeed, dolphins avoided whalewatching vessels but there was no change in behaviour when ships, yachts or fishing boats were in the area. Corkeron (1995) showed whalewatching vessels affected the behaviour of humpback whales (*Megaptera novaeangliae*) migrating through Hervey Bay, Australia, especially in the case of whales with calves.

A more favourable response was reported by Ransom (1998), who analysed encounter duration of whalewatching vessels around spotted (*Stenella frontalis*) and bottlenose dolphins on Little Bahama Bank, Bahamas. Between 1996 and 1997, observations of boat interactions within 1.2 km of the tour vessel and of swimmer interactions in the water found significant increases in encounter duration. Possible explanations included: dolphin habituation to swimmers, dolphin tolerance of humans and increased operator experience. More recently, Bejder *et al.* (1999) showed Hector's dolphins (*Cephalorhynchus hectorii*) readily approached dolphin boats but their behaviour changed, with less frequent approaches, when encounters exceeded 70 minutes. Whalewatching vessels have been used successfully to assess the relative abundance and distribution of minke whales (Leaper *et al.*, 1997).

Monitoring studies have also been carried out from land. For example, Yin (1999) studied dusky dolphins (*Lagenorhynchus obscurus*) in Kaikoura, New Zealand, from land, and reported statistical differences in their behaviour, depending on vessel distance from the dolphin group. There was a tendency for more course changes by dolphins when vessels were within 100m and 300m of groups. Briggs (1991) described 24-hour observations conducted from blinds on shore at two main rubbing beaches for killer whales (*Orcinus orca*). He reported a 50% decrease in time at rubbing beaches from 1987 to 1989. There was a near constant presence of fishing vessels during the season and gun shots were heard on 35% of the days that fishing vessels were moored (directed at shore and in water). Rubbing beaches are considered such an important part of the whales' habitat that a 'no go zone' was established to protect them. Recreational boats following whales in straits were usually stopped from entering the reserve by a ranger.

Impacts of whalewatching are likely to be cumulative rather than catastrophic, which emphasises a need for long-term studies and for cautious interpretation when evaluating disturbance from short-term studies (Bejder *et al.*, 1999). In addition, present information on baseline parameters is considered insufficient to measure subtle changes in behaviour that may be caused by whalewatching.

Long-term studies

Studies using long-term data sets are scarce. Watkins (1986) reviewed research cruise logbooks from whalewatching vessels in Cape Cod, Massachusetts, USA, over 30 years, to evaluate responses of whales to research vessels in the vicinity of Cape Cod. He compared whale behaviour before and after initiation of whalewatching and found whales responded primarily to underwater sound, light reflectivity and unexpected tactile sensation. The rate of habituation was often rapid, but varied with individuals and stimulus, and different species had different responses to vessels. He suggested that changes in whale behaviour have been gradual and therefore emphasised the need for long-term monitoring. Interestingly, although he recommended that skippers of whalewatch vessels use quiet, cautious approaches to whales, skippers felt that it was not necessary, as whales have apparently begun to accept the presence of whalewatch vessels. There are a number of long-term population studies using photo-identification from whalewatching vessels (IWC, 1990) and these data could be used to address the long-term impact of whalewatching. However, to date, no published review is available.

One novel attempt to monitor the effects of whalewatching involves Rhythm Based Communication (http://www.whalecontact.com/research) which attempts to determine the biological stress of the relevant individual(s).

The deficiencies in long-term monitoring have been recognised by both the scientific community and the industry, and there are a number of initiatives attempting to develop ongoing monitoring programmes (e.g. http://www.planeta.com/planeta/01/0103whales.html; Berrow & Holmes, 1999).

Monitoring visitor satisfaction

Managing whalewatching is as much about managing people as it is managing whales (Orams, 2000). In order to develop sustainable ecotourism, the monitoring of people and product satisfaction is also essential. However, in spite of the economic importance of whalewatching, there have been few surveys to determine whether or nor whalewatching operations are sustainable, and even fewer to assess whether the needs of those who will pay to see whales are being met (Orams, 2000).

The presence of whales and their proximity to the boat clearly influences the whalewatchers' satisfaction but Orams (2000) showed that a high degree of customer satisfaction can also be achieved in the absence of whales. It is important to note that there are a number of factors other than whales for successful whalewatching, and identifying and providing these elements is as important as watching the whales.

Tour operators are becoming increasingly aware of the necessity to adopt good practices and ensure that impact, both environmentally and culturally, is minimised (see http://www.toinitiative.org/home.htm). Sustainable management of whalewatching in many countries is being driven by tourism agencies and organisations, as well as by conservation bodies and a partnership between both is essential for successful implementation.

Requirements of monitoring programmes

A clear indication of the objectives of a monitoring programme is essential, as different indices will monitor different aspects of the life history or habitat of a species. For example, the EU Habitats Directive requires a 'favourable conservation status' to be maintained in designated MPAs, and monitoring programmes should be designed to determine compliance with this objective.

A long-term ongoing monitoring scheme should measure parameters that are sensitive enough to detect change at the appropriate scale. Analysis of these data may act as an early warning that something is changing, which may be an indication that the target species is receiving too much attention, and this should trigger a dedicated study. IFAW *et al.* (1995) provide a list of potential biological and operational parameters that could be used to monitor impact. Some parameters will be more useful than others and are species or location specific.

Long-term monitoring must also be financially sustainable: yet attempting to monitor population changes through recruitment, mortality or immigration/emigration is likely to be financially unsustainable (Wilson *et al.*, 1999). Monitoring will be more effective if a regular commitment is maintained over a long period, providing extensive reporting, rather than short-term, intensive studies.

Summary and Recommendations

The main recommendations from this work concerning the frameworks and monitoring required to develop genuinely sustainable whalewatching can be summarised as follows:

- Relevant authorities must develop the capacity to regulate whalewatching through licenses, as the most successful management potential exists where the number of operators and vessels are licensed.
- MPAs are most relevant where critical habitats or resources have been identified. MPAs require adaptive management, that is a form of management based on solid scientific grounds, and on performance studies, allowing a systematic evaluation to be made of the degree to which management objectives have been attained.
- Models of best practice, including codes of conduct and accreditation schemes, should be developed and promoted for the management of whalewatching.
- Research should be an essential element in the sustainable management of whalewatching, due to the lack of basic information on the ecology of cetaceans and the impacts of tourism. Research should not be seen as having a negative impact on whalewatching, as many whalewatchers are willing to pay more if some of the tour proceeds go towards whale research or public education.
- Education and interpretation should be an integral part of developing sustainable whalewatching.
- The involvement of stakeholders in resource management is critical to the success of sustainable development.
- Funding of monitoring programmes should be met by operators and considered an operating overhead, similar to boat fuel and insurance, and could be sourced through a levy system.
- In order to develop genuinely sustainable whalewatching, the effect of tourism activity on the species and habitat being exploited must be quantified and the impact assessed.
- Monitoring people and product satisfaction is also essential to the development of sustainable ecotourism.
- Long-term monitoring schemes should measure parameters that are sensitive enough to detect change at the appropriate scale and must be financially sustainable.

Acknowledgements

This work was carried out as part of the transnational research project called Marine Ecotourism for the Atlantic Area (META-) (Interreg: EA-C1UK3.14). Many thanks to Simon Allen, Peter Beamish, Lars Bejder, Dorete Bloch, Robert Bowman, Eduardo Dominguez, Erich Hoyt, Wiebke Finkler, Alexandros Frantzis, Jonathan

Gordon, Maddalena Jahoda, Sara Magalhaes, Mark Orams, Chris Parsons, Erika Urquiola Pascual, Christoph Richter, Fabien Ritter, Gerry Sanger, Regina Silvia, Alex Wilson for supplying information used in this report, Dr Erlet Cater for making comments on an earlier version of this chapter and David Bruce, Julie Wilson and the META- team for all their help and good company.

References

Au, W.W. and Green, M. (2000) Acoustic interaction of humpback whales and whalewatching boats. *Marine Environmental Research* 49, 469–81.

Bejder, L., Dawson, S.M. and Harraway, J.A. (1999) Responses by Hector's dolphins to boats and swimmers in Porpoise Bay, New Zealand. *Marine Mammal Science* 15, 738–50.

Berrow, S.D. and Holmes, B. (1999) Tour boats and dolphins: Quantifying the activities of whalewatching boats in the Shannon estuary, Ireland. *Journal of Cetacean Research and Management* 1 (2), 199–204.

Berrow, S.D., Holmes, B. and Kiely, O. (1996) Distribution and abundance of bottle-nosed dolphins *Tursiops truncatus* (Montagu) in the Shannon estuary, Ireland. *Biology and Environment. Proceedings of the Royal Irish Academy* 96B (1), 1–9.

Blane, J.M. and Jaakson, R. (1994) The impact of ecotourism boats on the St Lawrence beluga whales. *Environmental Conservation* 21, 267–69.

Briggs, D. (1991) Impact on killer whales: Impact of human activities on killer whales at the rubbing beaches in the Robson Bight Ecological Reserve and adjacent waters during the summers of 1987 and 1989. Report to Ministry of Parks, British Colombia.

Carlson, C. (2000) A review of whalewatching guidelines and regulations around the world. International Fund for Animal Welfare. Report to the International Whaling Commission SC52/WW5.

Corkeron, P. (1995) Humpback whales (*Megaptera novaengliae*) in Hervey Bay, Queensland: Behaviour and responses to whalewatching vessels. *Canadian Journal of Zoology* 73 (7), 1290–99.

Duffus, D.A. (1988). Non-consumptive use and management of cetaceans in British Columbia coastal waters. Unpublished PhD dissertation, University of Victoria, Victoria, B.C., Canada.

Duffus, D.A. and Deardon, P. (1990) Non-consumptive wildlife-orientated recreation: A conceptual framework. *Biological Conservation* 53 (1), 213–31.

Findlay, K.P. (1997). Attitudes and expenditures of whale watchers in Hermanus, South Africa. *South African Journal of Wildlife Research* 27 (2), 57–62.

Gordon. J.C.D., Leaper, R., Hartley, F.G. and Chappell, O. (1992) Effects of whalewatching vessel on the surface and underwater acoustic behaviour of sperm whales off Kaikoura, New Zealand. New Zealand Department of Conservation, Science and Research Series No. 32. Wellington: NZ Department of Conservation.

Hoyt, E. (2001) Whalewatching 2001, Worldwide tourism numbers, expenditures and expanding socioeconomic benefits. Yarmouth Port: International Fund for Animal Welfare.

IFAW *et al.* (1995) Report of the Workshop on the Scientific Aspects of Managing Whalewatching, Montecastello di Vibio, Italy.

IFAW (1996) Report of the Workshop on the Special Aspects of Watching Sperm Whales, Roseau, Commonwealth of Dominica.

IWC (1990) Individual recognition of cetaceans: Use of photo-identification and other techniques to estimate population parameters. Special Issue 12, *IWC Reports*. Cambridge: IWC.

IWC (1994) Report of the Forty-Fifth Annual Meeting, Appendix 9. IWC Resolution on Whalewatching. *Report of the International Whaling Commission* 44, 33–4.

IWC (1995) Report of the Working Group on whalewatching. *Report of the International Whaling Commission* 45, 32–3.

Janik, V.M. and Thompson. P. (1996). Changes in the surfacing patterns of bottlenose dolphins in response to boat traffic. *Marine Mammal Science* 12 (4), 597–602.

Kelleher, G. and Kenchington, R. (1992) Guidelines for establishing Marine Protected Areas. A marine conservation and development report. Gland, Switzerland: IUCN.

Kelly, J.E. (1983). The value of whalewatching. Unpublished paper, Whales Alive Conference, Boston, 7–11 June, 1983.

Leaper, R., Fairburns, R., Gordon, J., Hiby, A., Lovell, P. and Papastavrou, V. (1997) Analysis of data collected from a whalewatching operation to assess the relative abundance and distribution of the minke whale (*Balaenoptera acutorostrata*) around the Isle of Mull, Scotland. *Report of the International Whaling Commission* 47, 505–11.

Orams, M. (2000) Tourists getting close to whale, is it what whalewatching is all about? *Tourism Management* 21 (6), 561–69.

Ransom, A.B. (1998) Vessel and human impact monitoring of the dolphins of Little Bahama Bank. MA Thesis, San Francisco State University.

Samuels, A., Bejder, L and Heinrich, S. (2000) *A Review of the Literature Pertaining to Swimming with Wild Dolphins*. Maryland, USA: Marine Mammal Commission.

Scheyvens, R. (1999) Ecotourism and the empowerment of local communities. *Tourism Management* 20 (2), 245–49.

Tilt, W. (1985) Whalewatching in California: Survey of knowledge and attitudes. Unpublished paper, Yale School of Forestry and Environmental Studies, New Haven, CT.

Watkins, W. (1986) Whale reactions to human activities in Cape Cod waters. *Marine Mammal Science* 2, 251–62.

Wilson, B., Hammond, P.S. and Thompson, P.M. (1999) Estimating size and assessing trends in a coastal bottlenose dolphin population. *Ecological Applications* 9 (1), 288–300.

Yin, S.E. (1999) Movement patterns, behaviours, and whistle sounds of dolphin groups off Kaikoura, New Zealand. MSc Thesis, Texas A & M University.

Chapter 5
A Methodology for the Determining the Recreational Carrying Capacity of Wetlands

MARÍA JOSÉ VIÑALS, MARYLAND MORANT, MOHAMED EL AYADI, LOLA TERUEL, SALVADOR HERRERA, SANTIAGO FLORES and OSCAR IROLDI

Introduction

This chapter focuses on a methodology for determining recreational carrying capacity with specific reference to wetland environments. The methodological approach has been devised from an analysis of management tools for sustainable tourism in rural and natural areas. Carrying capacity is a management tool for the public use of recreational areas, since it provides a basis for the sustainable use of resources. For this reason, it could be argued that it is a useful management tool for fragile areas that are promoted as tourist attractions.

The issue of carrying capacity has often been addressed with empirical approaches: usually a process of trial and error. In the scientific literature there are also proposals for calculating recreational carrying capacity, through the application of formulae that become impossible to extrapolate to other situations. The research that forms the basis of this chapter, meanwhile, comprised a sequence of phases for the analysis of carrying capacity (e.g. physical carrying capacity, actual carrying capacity and permissible carrying capacity). All components and factors that should be included in such an analysis were studied: the wetland environment to be considered, profile of users, resources involved, type of recreational activity, systems of indicators, quick assessments, etc. One of the most important objectives of the study was to produce a model that could be extrapolated to other sites and situations. Therefore, the chapter proposes a particular research methodology, as opposed to a generalised formula. The various possibilities offered by new technologies as a basis for managing the carrying capacity of wetland ecosystems are also discussed.

Wetlands and Recreational Activities

Tourism in natural areas poses important challenges to environmental and tourism planners, managers and those trying to make conservation of ecological and cultural values compatible with the public use of ecologically fragile areas. The increase in these forms of tourism is connected to the interest society has for the environment, the desire to enjoy environments other than the cities (which are becoming ever more similar to each other) and also perhaps more traditional tourism activities becoming exhausted.

Outdoor leisure in natural areas can be appreciated, to different degrees, by many different sectors of the population. Recreational activities permit rewarding leisure pursuits in many ways: physical (e.g. walks), aesthetic (e.g. landscape contemplation and natural beauty), intellectual (e.g. educational and vocational achievements such as the study of geology, biology, archaeology, history) and social (e.g. family reunions, outings by groups of friends). This interest in the natural environment does not appear to be a passing fancy in terms of tourism demand but a social phenomenon on the increase.

Wetlands have not traditionally been the scene of recreational activities or indeed of other prosperous economic activities, except perhaps salt extraction. However, there has been increasing interest in their value as a recreational resource. In the past, the prospect of using wetlands was often rejected due to their low profitability (in agricultural terms), to the difficulties of traffic access and (in certain countries) to their tendency to harbour malaria. Such limited knowledge and appreciation of these ecosystems has seriously affected their consideration as viable resources. This has often resulted in them being neglected, particularly in areas suffering strong economic and demographic pressure. Ironically, however, had the recreational potential of wetlands been recognised long ago, then the same kind of critical state might have been reached for other reasons.

Traditionally, wetland landscapes have not been considered as a resource for recreational activities and, indeed, the literature has always been prone to praising the virtues of other areas such as the mountains and woods, due to the romantic spirit they can inspire. Wetlands have always been privileged places for hunting waterfowl and for fishing. Nonetheless, it should be pointed out that such activities did not have a recreational orientation but rather they were an economic activity that returned a living for many families in the locality. These activities still persist, due to tradition, but in most wetlands they are no longer the mainstay of these families but have instead become sporting activities.

Fortunately, wetlands have begun to achieve greater international recognition, attributable to growing ecological interest in their biodiversity and, more recently, an appreciation of the cultural heritage that many of them harbour. In fact, in terms of biodiversity, wetlands are arguably one of the richest ecosystems in the world, which is why they are increasingly being considered as nature sanctuaries. They have also awakened great interest in the interpretation/educational forms of tourism (ecotourism, cultural tourism, etc.), headed for years by the 'birdwatchers'. In short, circumstances are changing, exemplified by this growing awareness of the recreational potential of wetland environments. Tourist awareness of wetlands has considerable implications, which is why environmental and tourism managers must address the ever-growing recreational needs of those who demand recreational access to these areas.

For their part, EU regional development strategies have had an influence on tourism in natural and rural areas. Through its rural development policy (retaining populations in rural and marginal areas of the Member States, for example, through the reform of the Common Agricultural Policy) and its conservation policy (for example the establishment of the Natura 2000 Network), the EU tries to encourage tourist activities in these areas. Their objective in doing so has been to compensate for diminishing agricultural income and to support the economic cost of maintaining European protected areas.

Being aware of all these circumstances, the World Tourism Organisation (WTO), together with the United Nations, has designated 2002 as the International Year of Ecotourism (IYE 2002). We hope that this will consolidate 'interpretative tourism' as an option for responsible access to nature and local populations and, at the same time, generate the economic flows needed to raise the standard of living in the host countries. As such, visitor management plans have become a necessary element in the management of natural areas. If the trend in previous decades was towards the drafting of resource-focused master plans for natural areas, then lately more emphasis is being given to visitor management programmes. Arguably, carrying capacity has emerged as one of the most useful tools with which to implement and monitor them.

The Concept of Carrying Capacity

Ingrained in the concept of carrying capacity is the idea that there is a limit to the number of visitors to a tourist destination and that this limit is closely linked to the features of the site. Should this limit be exceeded, it would result in the onset of degradation – a direct threat to the survival of the (attractive) resource. This is the essence of carrying capacity. The problem, however, lies in deciding how this limit to use can be defined and quantified.

The question of whether or not there are identifiable limits to tourist activity is not a new one. In fact, the carrying capacity concept was originally used in the cattle-grazing realm, to calculate the maximum number of heads a territory could sustain without endangering its reproductive capacity. The concept has also been used in ecological studies to determine the local capacity of habitats in terms of the levels of animal population that the ecosystem could sustain.

In the 1980s, the concept of the 'limits of acceptable change' (LAC) gave greater emphasis to the decision of *when* the changes to a site would become unacceptable

(Stankey & McCool, 1984). LAC also gave a predominant role to the capability of the competent administration to respond, in that carrying capacity would be the threshold of human activity beyond which environmental impact on the resources will result (Wolters, 1991).

Today, the concept of 'sustainability' is popular. For example, it was put forward in 1991 by the World Conservation Union (IUCN), the United Nations Environment Programme (UNEP) and the Worldwide Fund for Nature (WWF), who declared that the capacity of an ecosystem is what it can bear and still sustain healthy organisms while maintaining its productivity, adaptability and renewal capacity (IUCN/PNUMA/WWF, 1991). A question still remains, however, as to how healthy the organisms within the ecosystem must be in order to thrive, rather than merely to be sustained.

Sustainability comprises two main parts: resources and their use. In this way, reference to the use of the resources (a more economic perspective) concerns the maximum limit to which a population may grow and continue being sustained long term by the environment (i.e. limiting use of the resource) without degrading its dynamics or its functions. From the perspective of the resources themselves (a more environmental perspective), consideration is given to the distribution in time and space of the non-renewable resources and on the rational use of the renewable resources, as determined by their rate of reproduction.

Applications of the concept of sustainability to tourist activity tend towards achievement of a particular combination of number and type of visitors and the impact of these activities on the destination environment. In this way, the area in which these activities are carried out can be used without compromising environmental quality. Tourism development policies are, therefore, at a crossroads. On one side there are the economic pressures of entrepreneur and investor groups, the population's (often urgent) need of employment and the necessity of generating income for the region. On the other side, there is the conservation and protection of the natural and cultural heritage. Factors such as the natural resources, ecosystems and local populations tend to determine the use limits in this respect. Such limits can affect the dynamics that give the resource and human factors sufficient stability to remain in time and assure environmental sustainability, while maintaining the local customs and quality of life.

Advances are required in both research and the application of tools for desirable control and prevention of impacts stemming from human activity in the natural areas, to fulfil the conditions of sustainable development. 'Recreational carrying capacity' and 'tourism carrying capacity' are some such tools. Recreational carrying capacity is applied to what, in tourism terminology, are called 'tourist areas' (settings). These are part of much larger 'tourist destinations' that contain other different tourist areas (settings). A tourist area is one that has an attraction or appeal that enables a specific tourism activity to be carried out. The analysis of tourism carrying capacity is more appropriate for tourist destinations where, besides environmental and social aspects, economic factors also become part of the assessment equation.

Certainly the concept of tourism is much broader than that of recreational activity, which is what this chapter deals with. The former includes elements corresponding to facilities that are necessary for the development of the economic activity (hotels, restaurants, etc.). The latter refers to activities carried out in natural, rural and / or urban areas during leisure, time which can take place either outdoors or indoors. These can be grouped into three basic modalities: leisure, sports / adventures and interpretative / educational (Viñals, 1999). Another important element to be considered is 'tourism' or 'excursionism' carrying capacity. The two differ in the use of basic facilities in that tourists sleep at the destination and use the hotel facilities and installations, while excursionists are on a one-day visit.

Assessment of tourism carrying capacity goes back to the 1970s when the UK National Trust spoke of the carrying capacity of recreational activity recognising that there must be limits that, should they be exceeded, would result in an undesirable degradation of the resource or attraction. Ceballos-Lascuraín (1998) recognised that tourist carrying capacity is a specific form of environmental carrying capacity. These definitions were somewhat vague and were extended to include aspects concerning visitor psychology and facilities. Thus, Pearce and Kirk (1986) argued that tourist carrying capacity also comprises 'visitor' psychological or perceptual factors referring to satisfaction in the recreational experience, besides the environmental factors. WTO (1993) added the socio-economic factor, considering (cf. McIntyre *et al.*, 1993) a threshold of maximum recreational use that, if exceeded, would give way to adverse effects on the society, economy or culture of the destination, which clearly enter into the concept of tourist carrying capacity. However, to consider and assess these psychological and socio-cultural factors, it is also necessary to resort to subjective parameters of evaluation (tourist behaviour patterns, design and management of facilities, customs, etc.), factors that are often difficult to pin down and quantify.

There are many international and national sources on sustainable recreational activities in natural areas, although they do not specify methodologies for the study of carrying capacity, rather they refer to sustainable tourism in general. Carrying capacity is usually regulated according to economic, administrative, functional or environmental criteria. However, the cases to which overall approaches that efficiently combine these criteria have been applied are few indeed.

On the specific subject of coastal zone management, some EU-level documentation exists that occasionally refers to wetlands in explicit terms. The main reference document is the 'Demonstration Programme for Assessment of Coastal Zones', published in 1996, which specified the practical conditions that must be considered to achieve sustainable development in all the different situations of the European coastline, providing a framework of coordinated action between EU Member States. Concerning this programme, the European Community published a document for consideration in 1999, entitled 'General Principles and Political Options', with the aim of opening up discussion for the preparation a draft European strategy for the integrated management of coastal zones.

The European Council has issued some recommendations on the subject. To be noted in particular is one by the Committee of Ministers of Member States, No. R(97)9, which relates to the development of sustainable tourism in respect of the environment in coastal zones. Staying with this subject, UNESCO has also published 'Coastal Areas. Managing Complex Ecosystems' in its dossier collection 'Environnement et Développement'.

In the Mediterranean region, for example, there is a programme called 'MedWet Coast', under the leadership of the wetland centre at Tour du Valat and framed within the MedWet Initiative (Mediterranean Wetlands). It is devoted to the coastal areas where management programmes, including aspects dealing with tourism in wetlands, are being developed. Its scope extends to wetlands in different countries of the Mediterranean basin, particularly along the North of Africa and Middle East.

Many specific initiatives can be observed in terms of the management of sustainable tourism and the development of tools for this task but all of these have been carried out through empirical methods, with varying degrees of success, and none has been applied in other situations. For this reason, the development of management tools such as carrying capacity, using standardised methodologies to allow the transfer of the tools to any wetland, is both necessary and long overdue. With such tools, it may be possible to assess and regulate tourist and recreational activities and control the pressure they exert on wetlands, while conserving these natural resources.

The evolution of the concept of recreational carrying capacity has, from the beginning, had the objective of maintaining the quality of the resource. This has been concurrent with the quest for thresholds to alert managers to environmental degradation of the tourist resource. In addition, more factors have been incorporated with the effect that we are now faced with a concept that, while easily understood in principle, is very difficult to define, assess and evaluate. Such complexity, however, does not appear to have prevented the extension of the use of carrying capacity, as previously mentioned and, in the best of cases, it is applied moderately in different management contexts.

It would be logical to think that, given the increasing interest in determining growth thresholds, the carrying capacity concept would already be in a mature stage and hence, that we could easily find a clear definition and methodology for its application. However, although there are references to it in research and policy arenas, there persists a lack of uniformity in its definition, assessment and evaluation. In any case, many experts agree that it is a good tool to control resource degradation situations and to enhance the quality of life of local populations as well as the quality of recreational experience.

Recreational Carrying Capacity: Phases of Analysis

In approaching the recreational carrying capacity of a wetland or any other ecosystem, the analysis of the key elements and the setting concerned must first be determined. The carrying capacity may be analysed in three consecutive phases, beginning with the 'physical carrying capacity' or the simple ratio of 'available space' to 'mean visitor space'. Following this, the 'actual carrying capacity' is determined by weighting the physical carrying capacity with correction/reduction factors that are site-specific. These factors are obtained by considering the physical, environmental, ecological, social and psychological factors that might modify the condition and sensitivity of the resources. The third level corresponds to the 'permissible carrying capacity', that establishes the acceptable limit of use in considering the management capacity of the administration of each site. This can change should the capacity of administrative management be modified or as a result of a change in conditions, such as the building of recreational and/or tourist facilities or increasing numbers of personnel.

Key elements of analysis: definition of the recreational setting

The key elements of carrying capacity are as follows.

The geographic space

Traditionally, carrying capacity studies of spaces have been limited to a few types of area: longitudinal spaces, specifically-designed spaces (trails, etc.), private property spaces (cattle grazing pastures etc.), species-specific habitats, and architectural and urban spaces (historical centres, etc.). When defining a geographic space, there are basically two types – open and closed – and both can be placed in three territorial realms – natural, rural and urban. Indoor spaces present some limitations of use due to the limited capacity of their design: hence it is simpler to determine their use threshold. In these cases, determining the carrying capacity will mostly concern safety and perceptual aspects of the visit, which will be defined by visitor expectations of the quality of the recreational activity experience. Some visitors are satisfied with a cursory visit; others may desire a detailed interpretation and comprehensive tour; other sites attract a large number of in-depth scholars (ICOMOS/WTO, 1993). For outdoor spaces, the analysis is extraordinarily more complex, since the setting is often larger and more diverse, ranging from specific sites (viewing points, trails, etc.) to great expanses of terrain (marshes, lagoons, etc.). In this respect, all recreational facilities must be considered, even if they are artificially built (picnic areas, hides, etc.), as these can be overlooked in assessments of carrying capacity.

In the wetland recreational setting, nearly all activities will be in the natural outdoor setting, excepting those that take place inside a building or facility (interpretative centres, eco-museums, nature classrooms, historic buildings, hides, etc.). Determining the study area is the first stage in approaching the carrying capacity analysis, since this conditions the spatial scale of the work. Normally, any analysis of a *tourist destination* is done as a whole, from a generic perspective, while *tourist areas* within the destination demand more specific analyses. For generic determinations, emphasis is on the evaluation of the ideal balance in the setting, relative to tourists and excursionists. This can be observed in the work of van der Borg and Gotti (1995) in the historical centre of the city of Venice (Italy) and other cities, where they applied linear programming to determine tourist carrying capacity.

Resources

The resources element is also fundamental in determining carrying capacity and it should be kept in mind when analysing the main services the resources offer the tourist activity, i.e. the appeal factor, the support of the activity and the reception of wastes. From the outset, special attention should be paid to the fragility of the resource, as well as to its vulnerability. The main attractions of wetlands, other than the landscape, are its animals and plants, as well as the presence of important cultural heritage and ethnological resources.

Indiscriminate development of certain tourist areas has often led to the degradation of the overall appeal of the resource and landscape for tourism use. Such a situation supposes that one could stop along the way to revise the state of things, but it is generally difficult to deny the right to enjoy nature and to forsake the income to the community that is likely to be generated by any form of tourism.

Activity (or tourist mode)

In order to establish carrying capacity, it is necessary to know how the territory and its resources might be affected by the tourist activity. To do this, it is necessary to prepare inventories and analyse the most typical recreational activities (in terms of activities and corresponding facilities) in these type of spaces and the impacts they generate on the environment (Table 5.1).

In wetlands, most activities could be included in the interpretative/educational group. These tend to be the least damaging to the environment, the most respectful to the local population and, therefore, closer to the concept of sustainability. However, it should be recalled that in wetlands, activities such as hunting and fishing are still practised as sports in some places. In addition, it must not be forgotten that the demand for recreational activities that do not require specific skills or expertise is ever greater. To cater to these markets, spaces must be idenitified specifically so that they do not interfere with other activity types (ecotourism, cultural tourism, sports, etc.). If it is necessary to set up all these activities, zoning the area or staggering participation times should be considered.

Users

It is also necessary to study the profile of the users to establish the level to which their impact can affect the environment. The nature of the study is dependent on the tourism form in question. The visitor profile can be determined through surveys and the direct observation of behaviour.

Physical Carrying Capacity

Physical carrying capacity can be established by combining the 'available space' and 'mean visitor space' each tourism form requires, according to the psychological needs of those participating in it. When a particular ecosystem is evaluated, two aspects are fundamental. First, the 'useful surface for recreational activities' must be identified. This is analysed through photo-interpretation of satellite images and

Table 5.1 Main impacts affecting wetlands, derived from recreational and tourism activities

Activity	*Effects on the wetland*	*Corrective measures*
Water		
Pumping aquifers and detouring flows Superficial water flows to supply water to recreation and/or tourist facilities	Drainage and wetland loss Changes in hydroperiod Salinisation of coastal aquifer by marine intrusion in littoral wetlands Soil compaction and induced subsidenceChanges in biocenosis	Prohibition of any work destined to drain the wetland Sustainable exploitation of groundwater Restoration measures
Artificial reservoir of water	Changes in the wetland morphology and hydrology	Restoration measures
Sewage from recreational and/or tourist facilities	Eutrophication Salinisation Microbiological pollution	Application of waste water treatment plan
Construction of hydraulic facilities and others to develop recreational and tourist activities	Decrease of wet surface Decrease in rate of water infiltration in the aquifer Increase in drainage speed of the wetland Increase in superficial runoff and therefore, or erosion	Maintenance of traditional channels with natural bottoms Development of facility plans avoiding wetlands
Geomorphological formations		
In-filling (to urbanise or introduce facilities for recreational and/or tourist activities)	Destruction of morphology, vegetation and habitats Reduction of recharging function of aquifers Indiscriminate increase of access to the wetland Decrease of environmental humidity and precipitation Wetland loss	Absolute prohibition of practices that lead to in-filling of wetlands
Hydraulic and communication, facilities, etc. in the wetland	Fragmentation of the territory	Planning and designing facilities avoiding wetlands
In-fills with rubble or other solid wastes from the tourist activities	Disappearances of seepage and water tables Increase of (lixiviates pollution risks of sediments and soils)	Collection of inert materials and cleaning dumps Control of illegal dumps Prohibition of spills

Table 5.1 (*cont.*)

Activity	*Effects on the wetland*	*Corrective measures*
Fauna		
Excessive hunting pressure	General detriment of waterfowl Change of natural distribution of speciesLead pollution	Implementation of plans to regulate hunting
Excessive fishing pressure from sports fishermen or use of inappropriate methods	General detriment of fish populations and particularly of endangered species or those threatened of extinction	Implementation of sports fishing plans
Hydraulic and communications facilities, etc. in the wetland	Isolating the biocenosis	Planning and designing facilities avoiding wetlands
Introduction of exotic species	Detriment to autochthonous species and loss of biodiversity	Strict control and prohibition on introduction of exotic species
Flora		
Hydraulic and communications facilities, etc.	Local loss of vegetal mass	Planning facilities avoiding wetlands
Introduction of exotic species	Detriment to autochthonous species and loss of biodiversity	Strict control and prohibition on introduction of exotic species
Uncontrolled fires (chance and provoked) in the wetland	Indiscriminate loss of vegetal mass, habitat, biocenosis Decrease of precipitationIncrease of erosion	Promoting natural regeneration Vegetal restoration
Landscape		
Infillings	Degradation of landscape	Restoration measures
Urbanisation	Loss of landscape quality	Possible restoration measures in existing urbanisations
Presence of rubbish	Loss of landscape quality Bad smells Increase of sanitary threats	Collection of trash and clean-up of dumps

aerial photographs, supplemented by field work, and consists of zoning specific spaces, independently of ecological considerations, having the physical characteristics of the site where outdoor activities could be practised. That is to say, certain areas must be avoided by those undertaking the zoning, such as flood risk zones (infrared satellite images are very useful for detecting them), densely vegetated zones, etc. and, in general, all those surfaces having difficult access (especially if non-specialised recreational activities are under consideration). The result is a map of zones with different degrees of recreational activity possibilities, depending on the forms of tourism under consideration.

The second (but no less important) factor to be analysed relates to the psychological or perceptual components. This factor deals with the conditions under which the recreational experience will be carried out in terms of product quality. It is often very difficult to establish these parameters because they must be sought by collecting subjective data (which may not always be the same for all visitors, even for those interested in the same attraction). Thus, the visitors' need of space to carry out the recreational activity under adequate conditions, or the 'mean visitor space', must be determined. This allows the establishment of a maximum number of people that can simultaneously remain in a recreational activity area without significantly affecting the quality of their experience. There are little data available on space needs for some recreational modalities and obviously any estimation is subjective and must be obtained through enquiries and surveys of the visitors. It is often necessary to begin with a frequency analysis, since there are no reliable data on the number of people that visit natural areas, not even in the cases of protected areas that have moderately regulated access. In the case of wetlands, which have a shorter tradition of recreational activity than other ecosystems, the situation could be even more complex but, equally, it could be argued that the situation is simplified due to these traditional circumstances.

Visitors practising more interpretative tourism forms, such as ecotourism and cultural tourism, often have a very low perceptual capacity. This means they may not easily tolerate the presence of too many visitors, even if they are practising the same activity. Such visitors often have a very specific profile. However, school groups carrying out educational activities can comprise larger groups, the same as visitors included in leisure activities (walks, boat trips, etc.). As regards hunting and fishing, there are often tacit or explicit rules that regulate these activities, often determined by associations, which specify a number of annual licences or fishing and / or hunting posts.

Actual Carrying Capacity

Actual carrying capacity can be established once the correction / reduction factors have been applied to a situation. These relate to the appeal and support of the resource (including, in this latter case, the local population). The process imposes a series of limitations that will reduce the possibilities established in the previous phase (physical carrying capacity). This level is closely linked with the analysis of the physical–ecological, cultural (in this case, the wetland has an ethnological heritage) parameters, as well as the socio-demographic ones. From these data we can define pressure and status indicators for the environment to make an 'indicator system'.

The use of indicators to evaluate recreational carrying capacity is taken from ecological monitoring studies, in which they are used to provide timely alarm signals on the state of the environment and its resources. Many organisations and countries, such as the European Commission (EUROSTAT, 1998), the OECD (1991), the United Nations Economic Commission for Europe, etc., have already defined or are preparing their indicator system. In general, they are useful when managers need to make decisions (Jiliberto *et al.*, 1996) since they provide a synthetic image of the environmental problems, which assists in forming opinions about the need to limit visitor access to the areas in question. The application of indicator systems to tourism dates back to the end of the 1990s. If environmental factors are related to socio-cultural and economic ones, we can obtain not only statistics on the environment but also indicators of sustainability, as the World Tourism Organisation has proposed (WTO, 1997).

The application of the indicators proposed in this chapter is based on a combination of the typology proposed by the OECD (1991), which relates to the situational context ('state indicators', 'pressure indicators' and 'response indicators') and the specific subject to which they refer (physical–ecological and socio-demographic). Thus, to determine the actual carrying capacity, the following groups of indicators have been chosen:

Environmental and human environment state indicators

These describe the quality of the environment and resources (natural and cultural) associated with tourist exploitation processes (whether it is an aesthetic resource, a support resource and / or a waste management resource). They can only measure observable changes in the environment (for example, the presence of pollutants in the water, noise, degree of biodiversity, etc.) and the local communities. These indicators are related to the physical-ecological and socio-demographic parameters because they consider the alterations to the natural and human system as a direct consequence of the recreational activities.

The physical–ecological parameters centre on the description of the biotope and biocenosis, which, together with the landscape, are considered to be resources of great interest from the point of view of conservation as well as for tourism activity. These parameters reflect the relative sensitivity of the resource to impacts resulting from the recreational activities on the natural systems, derived from land-use processes, the use of natural resources and the generation and emission of solid, liquid and gaseous wastes. The socio-demographic parameters describe the human environment.

This part of the process must also consider state indicators, which refer to the cultural environment and where a wetland has important cultural heritage, these resources must be subjected to the same analysis. To gather this information on the resources, it is necessary to establish a 'sampling and monitoring network' that records data periodically. This can be done manually or automatically using new technology. Detection systems, such as small sensors, can be camouflaged and placed throughout the area. Specifically, there are two types of sensors. The first type determines the presence of the public and are useful for calculating, in real time or later (at the end of the day or season), the number of visitors to the site. The second type directly measures the degree of impacts: acoustic pollution (particularly useful for wetlands, since the birds in the wetlands are especially sensitive to noise), temperature alterations, excessive presence of CO_2, etc.

The advantage of such systems of sensors is that they measure what actually concerns us, for example the threshold of noise tolerable to birds. Also, this information can be centralised, the data being sent through a digital data network to a computer centre for processing. In territories of reduced dimensions with sufficient cover, the data transmission can be done through GSM (digital mobile telephone), establishing simple networks of devices periodically sending data through SMS (Short Message Service). It is effectively a type of low-cost terminal, with a centralised data management and alarm system. The processed data allow the changes produced by the human presence to be estimated in real time. In this way, the projected number of daily visits permitted in the future could be amended immediately, if necessary. The disadvantage is that they provide very local information that must be complemented by other data.

Pressure indicators

These reflect direct and indirect pressures that the environment and the human environment experience. They can detect the recreational activities that directly affect the environment and its resources (for example, the discharge of untreated wastes coming from a tourist facility, the construction rate of installations in a natural interest area, etc.) and local populations and their culture. Socio-demographic parameters support the pressure indicators, since they centre on the human activities that impose negative effects on the environment.

Examination of these indicators requires data from statistics on the activities. It also requires data gathered *in situ*, obtained through surveys or directly with sensors, to measure the frequency of visits, e.g. those based in infrared barriers that count each passing visitor (similar to the method of counting the public at the entrance of museums and libraries, etc.). Present work in progress is based on indirect measures, i.e. from the number of calls made from mobile telephones in the area (which can be obtained from the mobile telephone operators). This method of analysing the frequency of visits is designed especially for those areas where access control is very difficult because they are open spaces. In any case, such data should be triangulated with another visitor counting system to establish a satisfactory level of reliability.

Another matter to consider is the periodicity of data gathering. We have seen that certain data can be obtained in real time, which is ideal, while other data (such as that relating to geomorphological or hydrological processes, or socio-demographic factors) are collected once a certain period of time has elapsed or (as is the case with much statistical data) are only registered on an annual basis. As such, it would be convenient that besides daily and/or weekly evaluations (particularly in high season), specific times be fixed throughout the year for evaluating the overall situation.

The selection of the indicators should respond to the criteria of scientific validity, representation, sensitivity to change, reliability of data, relevance, comprehensibility, predictability, comparability, extrapolability, cost-efficiency (in terms of cost of data gathering and use of information). The general acceptable change thresholds (LAC), that can be expressed in quantitative (absolute or relative) or qualitative terms, can be fixed later.

Table 5.2 presents some of the indicators that are the most useful to implement in wetland environments, at least in the first stages of analysis. Their practical application will allow improvement to the system and rejection of those that are unacceptable due to their cost, or because the quality of information contributed is not sufficient.

Some indicators involve more information than others or, at least, more detailed information. The monitoring of these indicators is a fundamental task in the development of this methodology, since the evolution of the site will allow efficient and effective evaluation of a master plan and, in this way, implement corrections and redirection, if necessary, to improve deployment. In this way, we propose to that 'Rapid Ecotourism Assessment' be used. This methodology is adapted from the 'Rapid Ecological Assessment', as developed by the Nature Conservancy (1992) with the purpose of preparing inventories, and evaluating and monitoring biodiversity in places of the world where logistic and economic resources are scarce. We propose the use of a variation on this technique – instead of studying the biodiversity of an area, the 'Rapid Ecotourism Assessment' analyses the tourism system elements present in a natural area that can suffer or produce impacts on the natural or cultural elements of the area. This is a flexible and dynamic process used to obtain and apply ecological and tourist information more rapidly. It is very useful in those situations where no detailed information is available or where there are great time limitations.

It must also be emphasised that, in those territories where vast expanses of land necessitate many sampling stations and/or automatic sensors, it is necessary to have a tool to relate the data gathered back to the geographic location in which the sensor was placed. This can be done through the use of a Geographical Information System (GIS). GIS is a very useful tool, capable of answering to the following five general questions:

- Location: Where is . . . ?
- Condition: What state is the resource in . . . ?
- Trend: What has changed since . . . ?
- Distribution: What are the existing distribution patterns?
- Modelling: What happens if . . . ?

Due to the impossibility of extending a cable or optic fibre network, development of a digital network for the transmission of these data could operate with VSAT (Very Small Aperture Terminal) sensors: small satellite communications antennae with their own data terminal. This type of device is used for remote data transmission from geographically distant points to a central processor and is usually used in managing hydrographical basins or for meteorological forecasting.

This instrument, which can be run via the internet, can be also used for the tourism promotion of the wetland (with other instruments such as a reservation centre), besides offering information about other products with similar characteristics in different places in the region or in the world. This type of marketing is used to control a geographically dispersed demand for the same form of tourism (birdwatching, for example) and responds to spatial trends in tourist behaviour. Its advantage, with respect to the carrying capacity, is that the user can determine the state of saturation of the site in advance, on the dates foreseen to travel, and can choose some of the alternative locations that the system presents. Following these criteria, a research team is currently designing a GIS programme for the Albufera of Valencia (Spain) that presents other information under these premises, as well as the existing alternatives in other wetlands in the Valencia region, for the educational/interpretative forms of tourism.

Despite these propositions, it must be stated that the use of indicator systems is not very widespread, but it can achieve optimum results and should always have a clear scientific basis. At the same time, it has an explicit social and political content. In practice, it is not always possible to choose to implement all of the existing technologies nor can an over-ambitious system of indicators be put in operation. To make analysis easier in these cases, it is very useful to start by identifying the most vulnerable resources and establishing their limits of use and the appropriate system of indicators. Since these are the most sensitive elements of the system, the rest of the resources will present less of a problem in their conservation.

Permissible Carrying Capacity

This phase of the process centres on determining the permissible carrying capacity, established from implementing correction–reduction factors, as derived from the management capacity (strategies in economic terms and technical management) of the administrations in charge (local, regional, national or international) and of the accessibility of the society itself.

The development, consolidation or maintenance strategies of recreational activities in the wetlands can be analysed from the political-economic type 'response indicators' that basically reveal the social and political stress level (Table 5.2). They indicate the actions taken by society, authorities and entrepreneurs to solve the environmental and socio-demographic problems generated by tourism and recreational activities (for example, fitting waste treatment plants in a hotel, drafting rules for public use, etc.).

Table 5.2 Indicators relating to wetland recreational activities

Resource	*Status*	*Pressure*	*Response*
Physical–ecological			
Water	– Water quality (presence of pollutants in superficial and ground- waters)	– Number of motor boats – Number of hunting licenses (lead pollution) – Volume of untreated sewage from their own recreational activity installations	– Investments in treatment plants – Drafting and approval of pollutant regulation standards in recreational activities (lead in hunting, boats, etc.) – Number of sanctions – Number of environmental education campaigns
	– Piezometric levels	– Rate of use of superficial and groundwaters (for recreational and tourist activities around the wetland – Demographic growth rate – Urban growth rate	– Drafting and approval of rules on water extraction – Drafting, broadcasting and application of Good Practice Codes – Number of sanctions – Investments in recycling and re-use of water. – Investment in awareness programmes for residents and visitors concerning sustainable use of water – Number of R&D project applications and grants devoted to the subject – Number of environmental education campaigns
Air	Air quality	– Number of vehicles with access to the the wetland – Number of pollutant gas emission points around the wetland	– Control and restriction of vehicule traffic in the wetland measures – Traffic assessment measures – Creation and use of public transportation systems measures – Increase in 'bike trails' kilometres
	Noise level	– Number of vehicles with access to the wetland – Intensity of traffic along the routes around the wetland – Total number of visitors per day/season/year – Size and behaviour of groups of visitors – Number of motor boats	– Control and restriction of vehicle traffic in the wetland measures – Assesssment of traffic around the wetland – Creation and use of public transportation systems measures – Measures for controlling groups of visitors

Table 5.2 (*cont.*)

Resource	*Status*	*Pressure*	*Response*
Flora	– Flora biodiversity index – Vegetal cover index	– Rate of decrease of vegetal species (especially the ones protected, threatened or in danger of extinction) – Number of fires and % of surface lost in fires – Rubbish increase rate in the wetland – Total number of visitors per day / season / year – Size and behaviour of visitor groups – Frequency of visits to vulnerable areas – Number of vehicles with access to the wetland – Number of boats – Facilities construction rate (change in use of land)	– Zoning measures – Vehicle circulation restriction measures in vulnerable areas – Drafting and approval of rules to control and prohibit introduction of exotic species – Number of campaigns and investments in forestry restoring – Number of R&D project applications and grants devoted to the subject – Number of awareness campaigns for visitors and local population – Number of sanctions – Number of environmental volunteers – Measures for controlling groups of visitors – Number of waste collecting facilities – Number of jobs created (Rural and Forest Rangers, Forest firefighters, etc.) – % of increase of protected surface – Number of natural heritage appreciation campaigns
Fauna	– Faunal biodiversity index – Population size – Number of nesting birds	– Number of fishing and hunting licenses – Number of species caught and felled – Decrease rate of animal species (especially those protected, threatened or in danger of extinction) – Level of fragmentation of the natural area (different artificial barriers, facilities, etc.) – Total number of visitors per day / season / year	– Zoning measures – Circulation restrictions measures for people and vehicles in nursery areas – Drafting and approval of rules to control and prohibit introduction of exotic species – Number of campaigns and investments in recovery and reintroduction of autochthonous species – Number of R&D project applications and grants devoted to the subject – Creation and approval of rules regulating the different uses and recreational activities (hunting, fishing, etc.) – Awareness programmes for visitors and local population – Number of sanctions – Number of environmental volunteers – Measures for controlling groups of visitors – Number of jobs created (Rural and Forest Rangers, forest fire fighters, etc.) – % of increase of protected area – Number of natural heritage appreciation campaigns

Table 5.2 (*cont.*)

Resource	*Status*	*Pressure*	*Response*
Landscape	– % of artificial elements in middle ground – Presence of rubbish dispersed throughout the area – Presence of visual barriers	– Number of building licenses granted around the wetland – Total number of visitors per day / season / year – Size and behaviour of visitor groups	– Drafting and approval of rules regulating landscape protection – Number of sanctions – Investment in landscape protection and restoration measures – Investment in programmes to adapt buildings to the environment – Number of waste collection facilities – Number of complaints from visitors and residents on landscape state – Number of natural heritage appreciation campaigns
Geo-morph-ological formations and soils	–Topographical changes – Changes in wetland perimeter – Hydroperiod changes – Changes in soil profile	– Total number of visitors per day / season / year – Increase in number of facilities – Level of fragmentation of the natural area (different artificial barriers, facilities, etc.)	– Visitor control and management measures – % of increase of protected surface – Drafting and approval of rules regulating land protection and control – Investment in land protection and restoration measures – Number of R&D project applications and grants devoted to the subject – Number of natural heritage appreciation campaigns – Number of monitoring programmes applied
Cultural Resources			
Cultural Heritage	– Number of buildings and other devices of cultural value – Cultural heritage conservation status – Number of museums	– Loss rate of built heritage – Total number of visitors per day / season / year. – Size and behaviour of visitor groups – Changes of use of cultural heritage	– Visitor control and management measures. – Drafting and approval of rules regulating cultural heritage protection and control – Investment in cultural heritage conservation, recovery, protection and restoration measures – Number of R&D project applications and grants devoted to the subject – Number of natural heritage appreciation campaigns

Table 5.2 (*cont.*)

Resource	*Status*	*Pressure*	*Response*
Socio-demográphic			
Society	– Loss of social values and customs – Natural growth rate of the population – Population migration movements – Employment rate – % of jobs in the wetland (hotels, rural rangers, etc.). – % of companies devoted to tourism and recreational activities – Rate of visitors/ tourists	– Rate of service demand growth – Rate of direct and indirect job creation (related to recreational activities and tourism) – Increase in density of resident and visitor populations – Price increase index rate – Migratory movement rate – Hotel growth rate – Restaurant growth rate – Growth rate of companies devoted to recreational activities	– Degree of satisfaction of residents – Degree of satisfaction of visitors– Resident complaints – Visitor complaints – Volume and number of donations and sponsorships – Number of local NGOs operating in the area and number of members – Number of training courses given and number of participants – Number of awareness campaigns – Number of environmental education courses given and number of participants – Number of R&D project applications and grants devoted to the subject (EU LEADER projects, etc.) – Revision, adaptation and creation of urban assessment plans

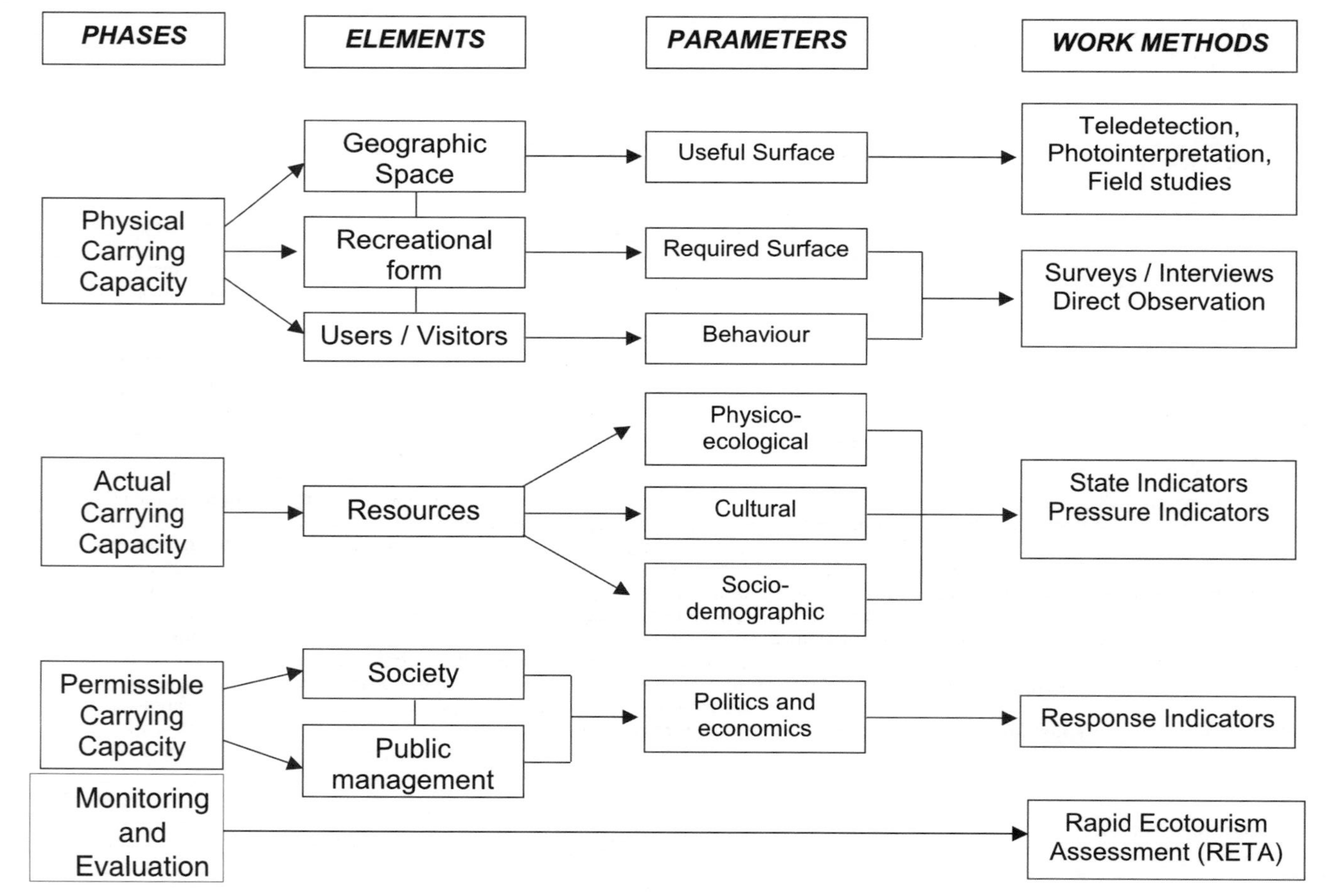

Figure 5.1 Procedure for applying recreational carrying capacity

Final Considerations

Implementing recreational activities in natural areas is a complex affair and environmental and tourism managers must learn to deal with initiatives and tourism strategies, considering both conservation and the growing trend of tourism and recreational demand. The management of visitors, as with resources, requires supporting technical, administrative and socio-economic tools. Recreational carrying capacity is presented in this chapter as a necessary technical tool. Arguably, the development method proposed here seems complex but this is not entirely true. It merely means systemising the available information and, if necessary, treating it in a simple manner (Figure 5.1).

It has not been the purpose of this chapter to identify a 'magic' number of visitors. This notion must be forsaken, due to the many factors that can condition the figure at any given point in time or space. Instead, we have tried to present all those elements and factors that should be taken into consideration in a natural area and, particularly, in a wetland, when planning and managing for sustainable tourism and recreational activities.

References

Ceballos-Lascuraín, H. (1998) *Ecoturismo*. Ed. Diana.

EUROSTAT (1998) *Indicadores de Desarrollo Sostenible*. Comunidades Europeas.

ICOMOS/WTO (1993) *Tourism at World Heritage Cultural Sites: The Site Manager's Handbook*. Madrid: WTO.

IUCN/PNUMA/WWF (1991) *Cuidar la tierra: Estrategia para el futuro de la vida*. IUCN/PNUMA/WWF.

Jiliberto, R., Mantelga, M. D., Sunyer, C., García Luna, M. M. and Álvarez-Arenas Bayo, M. (1996) *Indicadores ambientales. Una propuesta para España*. Ministerio de Medio Ambiente.

McIntyre, G., Hetherington, A. and Inskeep, E. (1993) *Desarrollo turístico sostenible. Guía para planificadores locales*. Madrid: WTO.

OECD (1991) *Environmental Indicators. A Preliminary Set*. Paris: OECD.

Pearce, D.G. and Kirk, R. (1986) Carrying capacities for coastal tourism. *Industry and Environment*, Jan.–Mar. 3–7.

Nature Conservancy (1992) *Evaluación Ecológica Rápida*. Programa de Ciencias para América Latina. Arlington, USA: Nature Conservancy.

Stankey, G.H. and McCool, S.F. (1984) Carrying capacity in recreational settings: Evolution, appraisal and application. *Leisure Sciences* 6 (4), 453–73.

van der Borg, J. and Gotti, G. (1995) Tourism and cities of art: The impact of tourism and visitors flow management in Aix-en-Provence, Amsterdam, Bruges, Florence, Oxford, Salzburg and Venice. UNESCO/Roste Technical Report No. 20.

Viñals, M.J. (1999) Los Espacios Naturales y Rurales. Los Nuevos Escenarios del Turismo Sostenible. In M. J. Viñals and Bernabe (eds) *Turismo en Espacios Naturales y Rurales* (pp. 15–33). Valencia: Universidad Politécnica de Valencia.

Wolters, T.M. (1991) *Tourism Carrying Capacity*. WTO/UNEP.

WTO (1993) *Indicators for the Sustainable Management of Tourism*. Informe del Grupo de Trabajo Internacional sobre Indicadores de Turismo Sostenible. Canada.

WTO (1997) *Guía práctica para el desarrollo y uso de indicadores de desarrollo sostenible*. Madrid: WTO.

Section 2

Experiences with Marine Ecotourism

Section 2
Experiences with Marine Ecotourism

While the purpose of the first section of this book was to raise issues about the nature, form and intended purpose of marine ecotourism, thereby providing a context for our study of the subject, the second section focuses broadly on stakeholder experiences with marine ecotourism. The intention is to assess the practice of marine ecotourism against its principles, to highlight areas of challenge and promise in the practice of marine ecotourism and to draw out some major practical lessons relating to the initiation, development, planning, management and regulation of marine ecotourism.

The section begins with Chapter 6, by Elizabeth A. Halpenny, which explores the role of non-government organisations (NGOs) in the conservation of natural and socio-cultural resources through their support of marine ecotourism. Six case studies from around the world are developed to demonstrate some of the positive actions that NGOs have taken in ensuring that marine ecotourism is effective in meeting conservation goals. Three major requirements for success are suggested: an appropriate setting, in which the natural and cultural resources required for marine ecotourism may be found and good access to major ecotourism markets may be secured; adequate resources in the form of the financial capital and skills required in order to make marine ecotourism possible; and appropriate planning and management of the marine ecotourism activity and the resource base upon which it depends.

The following two chapters focus on the management of diving tourism. In the first of these, Chapter 7, Ghazali Musa considers the development of dive tourism on the tiny Malaysian island of Sipadan. Musa reviews the impacts of diving tourism on the ecology of the island, arguing that diving activity has been for a variety of reasons grossly mismanaged in the past. Through a survey of diver satisfaction relating to a range of diving attributes and tourist impacts, the chapter identifies various management priorities for diving on the island. While divers seem for the most part to be highly satisfied with the quality of the diving experience available on Sipadan, this is thought to be mainly due to the richness of marine life, access to diving sites and friendliness and efficiency of the staff. Meanwhile divers expressed serious concerns with regard to the degree of crowding above and below the water, levels of noise and litter and poor underwater visibility. The chapter concludes that urgent action is required if unsustainable diving tourism is not to ruin this unique island ecosystem.

In the second chapter on diving tourism, Chapter 8, Claudia Townsend examines the potential role of education in addressing the environmental impacts of divers in the British Virgin Islands. The chapter reports on an experiment with modified dive briefings, which provides encouraging evidence that very simple educational tools can make critical differences to reef damage by tourists. Education is quick to develop and cheap to implement, and there is plenty of experience elsewhere to learn from. Indeed, substantial work has already been undertaken on how best to design education for tourists, and it is important that this experience is built upon and applied to the marine ecotourism context. Rather than attempting to reinvent the wheel, existing best practice must be modified for the marine ecotourism context.

Chapter 9, by Matthew McDonald and Stephen Wearing, also picks up on the theme of tourist education. The chapter examines the issues presently facing the Avoca Beach Rock Platform, Australia, and argues that the area would clearly benefit from the imposition of a framework, based on the principles of ecotourism, within which existing marine tourism might be managed. One of the central features of this management framework would need to be visitor education, particularly among line fishermen whose activities were found to be having significant negative environmental impacts on the area. The chapter goes on to argue that this could best be achieved through the development of a community-directed education programme, in which the community can learn to mobilise its resources, make decisions, develop a sense of ownership and take positive action on issues that are important to them. As such, this chapter crosses over between the themes of the education and interpretation component of marine ecotourism on the one hand, and local participation in marine ecotourism planning and management, on the other.

Community participation in the development of marine ecotourism forms the central focus of Chapter 10, by Zena Hoctor. The chapter argues that for marine ecotourism to achieve its overarching objective of sustainability, local participation should be established as a central feature of its planning and management. This implies involving the local community in all aspects of planning and management, at all levels, and from the outset. A case study of the development of 'Irrus', a marketing group for marine ecotourism in West Clare, Ireland, is then developed. The case study indicates how local participation in the development of marine ecotourism can best be encouraged and illustrates the benefits of the participatory planning approach in the development of genuinely sustainable marine ecotourism.

The next three chapters focus on practical issues in the development of marine ecotourism. Ilika Chakravarty, in Chapter 11, examines issues surrounding the development of a proposed marine park at Malvan, India. The study identifies a number of lessons for the development of ecotourism within marine parks, including the need for park development plans to deal adequately with the concerns voiced by local residents. For example, fishing is the major traditional occupation, yet the development plan does not make it clear that many such livelihoods are likely to be displaced by the establishment of the marine park. More and better information is needed on the number and type of alternative employment opportunities that are likely to be generated in the park, as well as the likely levels of income and taxation needed to address this concern effectively. The study also indicates that the social groups that are likely to benefit from the establishment of the park are likely to be very different to those who are likely to be displaced by it. The conclusion is that rather than merely being a desirable feature of marine ecotourism development, genuine local participation is critical to the success of such initiatives.

On a similar theme, in Chapter 12 Simon D. Berrow presents a case study of an attempt to develop a sustainable dolphin-watching industry in the Shannon estuary, Ireland. The Shannon estuary is the home of one of the few resident schools of bottlenose dolphins in Europe, and dolphin-watching trips into the estuary began on a modest scale in the mid-1990s. During the period 1999–2001, however, the demand for dolphin-watching trips increased significantly, emphasising the need for an effective system of management. A major move in this direction occurred in April 2000, when the area was formally designated as a candidate Special Area of Conservation (SAC) under the EU Habitats Directive. This enabled dolphin-watching to be made a notifiable activity, for the licensing of dolphin-watching vessels and for a code of conduct to be drawn up and implemented.

Chapter 13, by Colin D. Speedie, also considers experiences of developing sustainable marine ecotourism activities. The chapter discusses an initiative to determine the potential for the waters of Devon and Cornwall, in the South West of England, to support marine ecotourism. This involved a number of transect studies over the period 1999–2001, in which sightings of species that could potentially serves as targets for marine ecotourism operations were carefully recorded. While year-round marine ecotourism based entirely on cetaceans would appear to be entirely infeasible, opportunities do appear to exist for developing seasonal marine ecotourism around basking sharks, particularly if this is developed in conjunction with activities based on other marine and coastal species. Before this can become a reality, however, significant scientific research is required on the reproductive behaviour of basking sharks, their reactions to the presence of vessels or divers, and their vulnerability to collision with vessels.

Chapter 14, by Christos P. Petreas, considers the development of diving tourism in Greece and reviews the regulatory framework within which it presently operates. Scuba-diving is considered to hold significant potential as a sustainable form of tourism, that brings benefits to both the local communities in which it is based and to the environments in which it takes place. The chapter does, however, identify a number of areas in which existing regulation is insufficient to ensure the sustainable development of diving tourism. The chapter concludes that there is no reason why Greece cannot benefit from this form of tourism, provided that carrying capacities are identified and that regulatory measures are firmed up to ensure that such capacities are properly respected.

In this final chapter of this volume, Mark B. Orams, examines experiences of marine ecotourism in New Zealand, which is widely considered to be one of the world's foremost marine ecotourism destinations. He suggests that while tourism in New Zealand has always been based significantly on its 'natural wonders', there has been rapid growth in the development specifically of marine ecotourism in recent years. Marine wildlife, especially dolphins, seals, penguins and other seabirds, are particularly important attractions. Yet, perhaps in spite of its elevated position as a 'world class' marine ecotourism destination, the management of marine ecotourism in New Zealand presently faces a number of important challenges. These include the rapid growth of demand for marine ecotourism experiences, the adverse impacts of other economic activities on the marine environment, and the role of indigenous Maori people in the development of the industry.

Chapter 6
NGOs as Conservation Agents: Achieving Conservation through Marine Ecotourism

ELIZABETH A. HALPENNY

Introduction

This chapter explores the role of non-government organisations (NGOs) in the conservation of natural and cultural resources through their support for and use of marine ecotourism as a tool for positive change in coastal settings. Six case studies from developed and developing countries are used to highlight how NGOs have focused on four essential components of marine ecotourism:

(1) ecotourism-related financing mechanisms for conservation,
(2) the establishment of tourism industry and resource management standards and especially voluntary guidelines,
(3) research on the challenges facing the management of coastal and marine resources and marine ecotourism's ability to address these issues, and
(4) education of coastal stakeholders regarding solutions for coastal resource use problems including the implementation of genuine marine ecotourism.

As described in earlier chapters, marine ecotourism has inspired both hope and disillusionment through its use as an approach for achieving positive change in coastal settings. The principles established by META- (see Garrod *et al.*, 2001) and of the various industry codes of practice outlined in The International Ecotourism Society's (TIES) Marine Ecotourism Guidelines document (Halpenny, 2002) remain illusive, not surprisingly, due to their lofty and comprehensive nature. Some of the goals related to these principles include community development, biodiversity conservation, the safeguarding of cultural landscapes and traditions and the environmental education of tourists. Such goals are designed, in part, to generate a constituency of environmental and social sustainability advocates. Uneven achievement of success should not, however, discourage policy-makers and management practitioners from embracing marine ecotourism as a valued approach for achieving sustainable patterns of use and livelihoods in a coastal context. Indeed there are few other alternatives which show as much promise.

In appropriate settings, and with adequate resources and planning, marine ecotourism can be a success. Implementers of marine ecotourism continue to identify specific techniques and tools which can be used to fulfil marine ecotourism's mandate in coastal settings. While the fundamentals of marine ecotourism remain the same for each location, the approaches to implementing them vary. This chapter will investigate four essential elements of marine ecotourism – financing conservation, establishment of tourism and ecotourism standards, education and research – and outline how different NGOs have approached the implementation of marine ecotourism in specific settings. The issue of how these NGOs seek to achieve cultural and natural heritage conservation is emphasised. The case studies presented here, drawn on the author's experiences in both developed and developing countries, do not achieve all the principles of marine ecotourism. They instead serve as examples of how specific stakeholders have attempted to meet the needs of a specific coastal ecosystem, including its human inhabitants, through an experimental implementation of marine ecotourism.

NGOs and Conservation

Non-government organisations have emerged in the last decade as one of the principal advocates and implementers of marine ecotourism. Tourism and conservation NGOs can be membership organisations that have diverse memberships composed of community members, the tourism industry, conservationists, social activists and so on, or they may be specialised groups of stakeholders, for example environmentalists or consumer advocates. NGOs can also be non-membership associations.

NGOs have been criticised for voicing the concerns held by 'special interests' only, ignoring the broader public good. These views tended to be narrow in scope, their vision focused myopically on one setting, issue or stakeholder group need and displaying little understanding or desire to include the larger realities of an increasing globalised economy within their analysis. However, this phenomenon increasingly appears to be less evident as NGOs develop a mounting level of professionalism, going beyond simply criticising and being motivated by 'emotions and philosophies held by only a minority of the public' (Norse, 1993: 252). Many NGOs are moving forward with the agenda of sustainability, not only commenting on the debate about how to achieve sustainability but also developing projects of their own, experimenting with different approaches for achieving sustainability and conservation.

Conservation is the primary mandate of several international NGOs such as the Worldwide Fund for Nature (WWF) and the Wildlife Conservation Society, as well as many national and local NGOs. Many of these organisations have embraced ecotourism as a form of development that is complementary to the goals of conservation efforts. This is due in part to ecotourism's relatively modest negative impact on natural and, to a lesser extent, cultural environments, compared with other economic activities. It is also partly due to ecotourism's ability to provide an opportunity for economic benefits to local communities residing in the landscapes which conservation NGOs seek to conserve. International NGOs such as the Nature Conservancy and Conservation International demonstrate their belief in ecotourism as a development and conservation tool through the operation of ecotourism departments within their institutional frameworks. Understanding how NGOs use marine ecotourism to fulfil their conservation mandates and achieve positive change in coastal settings will be explored in the following sections, beginning with its role as a financing mechanism for conservation.

Marine Ecotourism and Financing Conservation

Increasingly, marine ecotourism is proving to be a valuable tool for financing the conservation of marine biodiversity and the traditions of coastal peoples. This is accomplished through the generation of revenue from marine ecotourism activities such as whalewatching, visitation of coastal museums and guided coastal hikes. Revenues directly benefit those involved in the marine ecotourism industry through the generation of income and the opportunity to operate tourism-related businesses. Indirect benefits such as the multiplier effects arising from purchasing products from non-tourism suppliers by tourism businesses also occur, although these positive impacts are usually quite small.

One specific interest of this chapter is the direct impact that marine ecotourism can have on the financial viability of Marine Protected Areas (MPAs): those parks that are designed to protect coastal and marine resources and seascapes. NGOs can play an important role in ensuring marine ecotourism provides an adequate economic incentive for its presence in MPAs. This is explored later.

Coastal areas are often heavily visited by tourists and MPAs are one of the leading destinations for both local and international tourists. MPAs are one of the key focal points for harnessing efforts to conserve and protect marine and coastal resources.

If tourists are using MPAs for part or all of their vacation needs, it follows that significant impacts arise from this visitation and the negative socio-cultural and environmental impacts have been well studied. However, less research has been devoted to understanding the potential positive benefits of tourism's use of MPAs, such as increased environmental awareness experienced by park visitors stemming from park-based interpretation programmes and economic support for communities in and nearby marine protected areas (Bookbinder *et al.*, 1998; CSQ, 1999; Dixon & van't Hof, 1997; Eagles, 2000; Goodwin *et al.*, 1997; Honey, 1999; Lindberg *et al.*, 1996; Marion & Farrel, 1998; McLaren, 1998; Smith, 1989; Walpole & Goodwin, 2000; Weaver, 1998).

Table 6.1 Selected MPA visitor fees

Australia	Great Barrier Marine Reef: \$4/day Ningaloo Marine Park: \$7.50/day (dive fee)
Belize	Hol Chan Marine Reserve: \$2.50/day Half Moon Caye: \$5/day Glover's Reef: \$5/day, \$20/week Lesser fees for Belizeans although fees are currently not enforced
Brazil	Abrolhos Marine National Park and Fernando de Noronha Marine Park: \$4.25/day
Costa Rica	Cocos Island: \$105/visit
Egypt	Ras Mohammed Marine Park: \$5/day foreigners and \$1.20 for Egyptians Red Sea Marine Park: \$2/day (dive), will increase at end of 2001 to \$5/day
Italy	Miramare Marine Reserve: \$2.20/day + fees for services e.g. dive trip is \$22, snorkelling is \$11
Netherlands Antilles	Bonaire: \$10/year (dive fee same for locals and foreigners) Saba: \$3/dive and \$3/week for snorkellers. Residents are not charged. Eustatius: \$12/year (dive fee) and \$12/night for yachts
Philippines	Tubbataha Marine Reserve: \$50 Gilutungan Marine Sanctuary: \$1/day for foreigners and \$0.50 for Filipinos

Source: Lindberg & Halpenny (2001).

The chief mandate of MPAs is conservation of marine and coastal biodiversity yet their ability to achieve this goal is severely curtailed by a lack of funding. A report by the Worldwide Fund for Nature (WWF) (2000: 1) supports this view, stating that most MPAs are 'under-resourced and poorly managed, offering little in the way of real protection. Global estimates suggest that as many as 70–80% of the MPAs that have been established worldwide are protected in name only and are not actively managed at all'. Government agencies, especially in developing countries, lack the funds to address this issue and alternative sources of revenue are needed. Many believe that tourism could be one of the answers to the funding problems of certain MPAs but little data have been collected to support or disprove this.

In a study of the current status of tourism-related financing mechanisms for MPAs, Lindberg and Halpenny (2001) found the use of fees for funding part or all of the operations of a MPA was becoming increasingly common in the marine conservation world. Fees also appeared to be used more often in MPAs than in terrestrial parks. The latter observation may be due to the fact that most MPAs are a relatively recent phenomenon, having become more common within the last 20 years, and have traditionally received less political and financial attention from the public and policy-makers. Examples of the fees charged by selected MPAs are listed in Table 6.1. On average MPAs generally charge entry fees ranging from US$1 to US$5 per day, or US$10 to US$20 per year.

NGOs have played a leading role in making visitor fees a successful option for funding MPAs. Two case studies, described here, help illuminate this role. In general, NGOs play two different roles: external NGOs, such as an international conservation NGO, help build coalitions of support for fee establishment or increase, while local NGOs serve as voices for particular interest groups, for example managers of a particular coastal destination or a coastal user group. These NGOs can work in tandem with government agencies and sometimes the tourism industry to produce an equitable and feasible fee system for parks.

Case study 1: Belize's MPA system – experimenting with fees

Several NGOs have been involved with recent efforts to increase and consolidate visitor fees in Belize's 12 marine parks. Belize has a long history of MPA fees; Hol Chan Marine Park began charging an entrance fee in the early 1990s. Since then, several other MPAs have been established, managed by a diverse mix of private sector associations, conservation NGOs and government agencies. This diversity of management agencies has produced a confusing and increasingly competitive system of parks, which make visitors who wish to see the largest barrier reef in the western hemisphere, pay multiple and differing fees for each park they visit. Park managers are increasingly interested in establishing and raising parks fees to take advantage of visitors' interest in Belize's coral reefs and other marine attractions such as mangrove forests, sport fishing shallows, whale sharks and so on.

Conservation NGOs such as the Programme for Belize and the Belize Audubon Society, the latter managing two MPAs itself, have been collaborating with government departments and ministries to research what visitors are willing to pay and what is the best approach to administering fees. An international NGO, The International Ecotourism Society, serves as an external advisor to the initiative, utilising its network of planning, management, industry and research professionals to document most recent fee efforts internationally and bring those lessons to Belize for assessment (Lindberg & Halpenny, 2001). Together, these agents are currently working on a pilot project that will implement fee changes and experiment with new fee payment schemes including the use of the internet for bulk ticket payments and administration.

Belize is a world leader in its efforts to manage revenue collected from tourists. Through its Protected Areas Conservation Trust (PACT), an environmental departure tax of approximately US$4 is collected from passengers at the airport and distributed to protected areas in the country in the form of grants for projects. PACT also derives some of its revenue from a percentage of park fees charged at Belize's protected areas (this is currently 20% but is scheduled to decrease to 10%). The Trust is an independent entity and while both NGOs and government agencies sit on the board of PACT, NGOs hold the majority of seats.

Case study 2: Introducing a visitor fee to Bunaken Marine Park, Indonesia

A second case study, from Indonesia, highlights how collaboration between an industry NGO, government agencies and conservation NGOs can accomplish change in a coastal setting that is under severe threat.

Bunaken National Marine Park, situated on the northern tip of the Indonesian province of Sulawesi, was established in 1991. The Park has rich biodiversity, including extensive mangrove forests and coral reefs. For years it suffered from a lack of funding, resulting in weak management and poor enforcement of protection laws. Meanwhile, dynamite and cyanide fishing threatened reefs and illegal forestry endangered mangroves. Several groups have worked together to establish a fee for visitors to the park. Local dive operators, represented by the North Sulawesi Watersport Association (NSWA), were very supportive of the initiative and were involved from the inception of the project. NSWA worked with park managers and government agencies such as Indonesia's National Resource Management Programme, Indonesian-based community and conservation NGOs and an international NGO.

As part of the project's preparations, a willingness-to-pay survey was conducted. The management of the collected fee was the chief concern for a majority of respondents. Visitors wanted to see the revenue go toward conservation programmes in the Park, rather than into the coffers of the government or the pockets of local officials. To address this issue, a pilot project was proposed for Bunaken; the government was lobbied for the creation of a more decentralised approach to fees management. The 'dive industry was a key ally in lobbying the government to pass the law' (Erdmann, pers. comm.) that would change how the fee revenue would be distributed. The Bunaken National Park Management Advisory Board (a multi-stakeholder board consisting of representatives from the dive industry, environmental NGOs, academia, villagers from within the Park and government officials) was created and receives 80% of the fee revenue, the other 20% being split between the national, provincial and two district governments.

The fee was developed and initiated over a ten-month period and came into effect in March 2001. Indonesian visitors pay a fee of Rp. 2500 (US$0.30) and foreign visitors (comprising divers, snorkellers and backpackers) pay Rp. 75,000 (US$8). Residents within the Park are exempted. The managers and board chose to introduce a relatively low fee for the first year for several reasons:

(1) to minimise industry and especially backpacker opposition,
(2) to prevent government officials from 'eyeing' the funds collected as a treasure trove to delve into, and
(3) to 'prove' to tourists that their fees are really doing something before asking for a larger fee – by starting small, they could avoid overly high expectations from tourists.

The managers and board estimate that it will require approximately $250,000 per year at a minimum to manage the park. Given current estimates of approximately 10,000 visitors, this would mean an eventual fee increase to US$25 per annum. The system is based on the Caribbean-based Bonaire Marine Park's model, in that when a visitor pays his or her fee at one of two entrance gates within the park or to a dive operator or travel agent (who buys passes in bulk from the Bunaken National Park Management Advisory Board), they receive a waterproof entrance tag which must be worn. As in Bonaire, the tag has become a collector's item. Indonesian day visitors receive paper tickets, as with other national parks. The implementation of the fees has gone very well and divers and dive operators are very supportive. Plans are now underway to increase the park fee in small increments, giving sufficient warning for industry stakeholders to adjust their pricing schedules (NSWA and Erdmann, pers. comm.).

While efforts associated with these two case studies are still underway, the efforts to pay for conservation through tourism appear to be possible, although still far from fulfilled. There are examples of MPAs that have achieved full financing of their operations through visitor fees and related tourism revenues such as souvenir sales and concessions. These included Ras Muhamed Marine Park in Egypt, Nelson's Dockyard National Park in Antigua and Bonaire Marine Park in The Netherlands Antilles, although this may not occur every year due to fluctuations in visitation and changes in economy, such as increased inflation (Halpenny, 2002). However, most parks cover only part of their operation expenses through tourism revenues. Despite this shortfall, tourism remains the best opportunity for parks in developing countries to fund biodiversity conservation, as other fundamental priorities such as education and health care continue to demand the attentions of these countries' small central treasuries. NGOs will continue to play an important role in ensuring park finances are improved through the presence of marine ecotourism. Not only can they help represent and resolve differing stakeholder interests and bring in outside technical expertise and experiences, they can also perform ongoing tasks such as running a 'Friends for the Marine Park' organisation, which sells souvenirs, or running guided interpretive walks for visitors as part of additional fund-raising efforts. All of these marine ecotourism activities must meet a certain level of excellence in service and experience, commensurate with regulations and guidelines designed to achieve sustainability.

Establishing Standards for Marine Ecotourism

Regulations, codes of conduct (for tourists) and codes of practice (for industry) are all part of larger efforts to set standards for conducting marine tourism and ecotourism in a given setting. While the term 'standard' can be used to denote specific and measurable indicators (for example, carbon emission levels), in this chapter it is being used as a general, 'catch-all' phrase to encompass the efforts of the tourism industry, as well as coastal managers to achieve a balance between use of marine and coastal resources and their preservation.

The principles outlined by Garrod *et al.* (2001) for marine ecotourism are part of this process, as are the codes of practice described in TIES' Marine Ecotourism Guidelines (Halpenny, 2002). More general codes of conduct and practice for ecotourism and tourism activities are also useful and can be found in documents such as Font and Buckley (2001), Tourism Queensland (2000) and UNEP (1998). NGOs have played a major role in the development of voluntary management initiatives and guidelines development, whereas regulations (enforceable by law) are more commonly associated with government agencies.

The two following case studies are used to illustrate how standards for marine ecotourism have been set. The first describes the methodology used to create The International Ecotourism Society's Marine Ecotourism Guidelines – generalised codes of practice that can be modified for individual settings and their associated challenges. The second describes a set of standards developed by the NGOs Rainforest Alliance and Conservacion y Desarrollo (Conservation and Development) for tour boat operations occurring in a specific setting, the Galapagos Islands. The second case study is interesting because of the inclusion of measurable indicators and their use to certify boat operators in the Galapagos. Both organisations and their guidelines have an underlying mission of ensuring that marine resources are conserved through the encouragement of tourism practices that are sustainable.

Case study 3: TIES' Marine Ecotourism Guidelines

This project began in the late 1990s, driven in part by the perceived importance of marine conservation and the lack of attention it had received to date. The International Ecotourism Society is a membership NGO, whose members are professionals, based in over 70 countries, and working in the field of ecotourism as planners, architects, researchers, community advocates, and so on. The Society had created two other sets of guidelines, one for nature tourism operators and a second for ecolodges (TIES, 1995; Mehta *et al.*, 2002). Lessons from the creation of these two documents were used to formulate the methodology for creating the Marine Ecotourism Guidelines.

While initial efforts such as the creation of an annotated bibliography by university partners in the first stages of the project helped frame the project's later efforts, the core strength of efforts to develop the guidelines was based on TIES' status as a membership organisation. Consultation with its members provided the richest sources of information and guidance for the Guidelines. This took three forms: the first was a mail-based survey of marine tourism operators, soliciting their comments on the role of guidelines in their region of the world, what coastal management issues were most important and what kinds of best practice they were employing. The second approach featured stakeholder meetings in the Caribbean, in which both members and non-members were encouraged to participate. The stakeholder meetings were organised by two Caribbean-based NGOs and one government agency. At these meetings, priorities that needed to be addressed in the Guidelines were suggested and best practice examples in the region were identified. The third form of membership involvement was the international review of the final document. Experts from many different disciplines, ranging from landscape architecture to human resource management, and based in a range of geographic locations, were asked for comments.

While international government organisations such as the World Tourism Organization, could conceivably undertake similar activities, these agencies are often under-funded and over politicised, and have chosen to ignore ecotourism as a focus. This highlights the importance of NGOs in this process (there have been exceptions to this, including UNEP's 1998 publication on ecolabels for tourism and current efforts associated with the United Nation's International Year of Ecotourism 2002). A second example of an NGO's efforts to set standards for marine ecotourism is illustrated in the following case study from Ecuador.

Case study 4: SmartVoyager Certified – creating a certification programme for Galapagos tour boats

Two NGOs, the Ecuadorean environmental group Conservation and Development (Conservacion y Desarrollo) and the US/Costa Rica-based Rainforest Alliance, collaborated with scientists, conservation experts, tour operators and others to set standards for the maintenance and operation of the tour boats in the Galapagos Islands, a world-renowned natural history destination. Through a consultative process, the NGOs guided the creation of standards that cover various potential sources of pollution (such as wastewater and fuels) and set rules for the management of everything from the docks to the small craft that ferry visitors ashore. Standards were also established for purchasing and supply, which are designed to minimise the introduction of alien species, and the requirement for good living conditions and advanced training for boat crew and guides. Finally, tourists who patronise these operators are given 'leave-no-trace' guidelines for their visits.

Specialists board the operators' boats to evaluate the vessels and measure their operations according to the standards. Five tour boats now display the SmartVoyager seal of approval (see Figure 6.1). Currently, efforts are underway to expand the use of the standards for other boat operations in Ecuadorean waters, especially for smaller tourism operators who face greater financial challenges in upgrading their vessels and operations to meet the tourism standards.

This second phase of the project is financed through World Bank money directed towards Ecuador. The SmartVoyager project was originally financed through

Figure 6.1 Smart Voyager Certification Seal

Table 6.2 SmartVoyager certification principles (standards updated February 2001)

Company policy: The company must have a management policy that includes compliance with national legislation and international agreements as well as SmartVoyager standards.
Conservation of natural ecosystems: The tourist operation must support and promote conservation of the Galapagos National Park and the Marine Reserve.
Reduction of negative environmental impacts: The tourist operation must prevent or mitigate and compensate for any environmental damage done to the Galapagos Islands and Marine Reserve.
Lowering the risk of introduction and dispersal of exotic species: The tourist operation must prevent the introduction of species from the continent to the islands and the dispersal of species between islands.
Just and proper treatment of workers: The tourist operation must elevate the socio-economic welfare and quality of life of the workers and their families.
Employee training: All personnel involved with the tourist operation must receive environmental education and training.
Community relations and local welfare: The company must make a commitment to the welfare and socio-economic development of the Galapagos Islands community.
Strict control of use, supply and storage of materials: Boat operators must plan and control the consumption, supply, and storage of materials, taking into consideration the well-being of tourists, workers, local communities, and conservation of natural ecosystems.
Integrated waste management: Boats must follow a waste-management plan, including reduction, reuse, recycling and adequate final treatment and disposal of all wastes.
Commitment on the part of the tourists: Tourists must be guided in their involvement in protecting natural resources and local cultures, tread lightly, and collaborate with the island conservation programs.
Planning and monitoring: Tourism operations must be planned, monitored and evaluated, taking into consideration technical, economic, social and environmental factors.

Source: Rainforest Alliance (2001).

unrestricted funds from both NGOs and incubated within an existing agricultural certification programme. It is difficult to find funding to address certification issues associated with tourism, and even more difficult to identify sustainable means for continuing their operation, as ecotourism operators are usually very small and cannot afford to pay fees large enough to finance a certification programme on an ongoing basis. Rainforest Alliance and Conservation and Development realise that a 'creative combination' of financing mechanisms including fees and grants, the latter based on the conservation value of the certification programme, is necessary.

The basic principles of this certification programme are illustrated in Table 6.2 (based on information provided by Rainforest Alliance, 2001; Jorge Peraza, pers. comm. and Ronald Sanabria, pers. comm.).

The creation of standards, especially those of a voluntary nature, has been a major focus for NGOs during the last decade. The key to the success of these standards, be they codes of practice or specific indicators used in certification, lies not only in their clarity and applicability to a given situation but also to the process of their development and the generation of a sense of 'ownership' among those stakeholders who must employ them. Locally-based NGOs and membership NGOs foster this form of acceptance and loyalty to the standards. An important part of this acceptance process is the generation of information through research and distribution of this information about the social and environmental challenges associated with marine ecotourism, ensuring that all stakeholders, including visitors, understand the need for such standards. The role of research and education will be explored in the following section.

Research and Education: Cornerstones of Marine Ecotourism and Conservation

Research and education are the final two roles played by NGOs in fostering conservation of marine resources through marine ecotourism to be discussed in this chapter. The two NGOs highlighted in the following case studies perform these activities very well; however, many NGOs tend to specialise in only one or the other.

Researching the promise and limitations of marine ecotourism to address conservation and the other objectives associated with sustainability (such as community development) is an essential role that NGOs have played. This is reflected in publications such as CSQ (1999), Honey (1999) and Mastny (2001). The literature features both positive and negative assessments of marine ecotourism's ability to accomplish change in a given setting. Research by NGOs also analyses the current status of marine resources and coastal communities' well-being, and has highlighted the need for alternative strategies, such as marine ecotourism, to solve coastal challenges.

This research is only useful if it is distributed to policy-makers, scientists, coastal managers and the general public. The role of educator played by NGOs, as part of this dissemination process, is an essential part of achieving conservation of natural and cultural heritage. These two case studies highlight how this is accomplished in Canada.

Case study 5: Group for Research and Education on Marine Mammals (GREMM)

Founded in 1985, the non-government organisation GREMM is located where the freshwater fjord river, the Saguenay, meets Canada's largest river, the St Lawrence. Located at this junction is the Saguenay–St Lawrence Marine Park. The Park has a unique co-management arrangement which is shared by the provincial government of Quebec and Canada's federal government. GREMM is responsible for much of the research and education that takes place within the Park. The NGO's mission is to understand and foster appreciation for the whales of the St Lawrence River and the marine ecosystem on which they depend. GREMM conducts scientific research on the whales and provides conservation education to tourists and residents. This knowledge encourages more informed choices, by helping to ensure the consideration of whales in decision-making processes.

Its education programme consists of a series of interpretation sites within the Park, the Center for the Interpretation of Marine Mammals – an award winning interpretation center in Tadoussac, Quebec – and a website dedicated to whales (www.baleinesendirect.net).

GREMM has also worked with whalewatching excursion boat operators and other stakeholders to establish a set of whalewatching guidelines suited for local conditions and the five species of whales that visit the region each year. One of GREMM's key factors in success has been a collaborative spirit and the use of partnerships in setting and achieving conservation goals.

Case study 6: Quebec Labrador Foundation (QLF)

QLF also promotes the use of partnership in its many projects located along the eastern coast of Canada and the New England coast of the United States. For more than 40 years, QLF has focused on community development, conservation of natural and cultural resources, environmental management, education and community health. Many of its projects are based in rural communities with little access to development capital or political power. The French Shore project serves as an example of how QFL tackles the issue of conservation through the development of marine ecotourism products.

The French Shore is located along the eastern portion of Newfoundland's Great Northern Peninsula. The region became connected to the rest of Canada by road just 20 years ago. Communities located in this region were traditionally involved in fishing and pulp forestry. Both industries have experienced significant declines in the last 15 to 20 years, with fishing suffering a near collapse with the disappearance of Newfoundland's chief fishery product, cod. The area, which is rich in cultural traditions associated with 'living off the land and sea' and dominated by Irish ancestors' music and foods, suffers continued neglect from regional governments. Unemployment is chronic and opportunities for self-advancement are rare.

The Quebec Labrador Foundation has worked in this region for many years and recently initiated a project designed to celebrate the tradition of French fishers who fished Newfoundland's shores for over 400 years, a tradition which ended in 1904. French fishers would come ashore for the summer in the French Shore region to process fish and prepare them for sale in Europe. Significant artifacts remain from this period, as do cultural heritage buildings associated with the more recent forestry and fishing industries established by Newfoundland settlers. Working with local government officials and community members, QLF is conducting research to document these treasures before they disappear. Visitor centres have been opened in each of the four communities in the French Shore to welcome tourists to the region and display the area's proud cultural heritage. Promotion of related ecotourism products in the visitor centers, such as whalewatching excursions and coastal hikes, is also carried out.

QLF also works to promote conservation of the region's heritage through the Traditional Skills Network, established to preserve local craft traditions through workshops and demonstrations at the visitor center. Over US$10,000 was raised in 2000 from the sale of quilts and other handcrafted items, a percentage of which goes to the craftsmen and women, with the rest being reinvested in the programme for the following year. Additional funding was also acquired from federal employment programmes.

In addition, a museum was opened in 2001, which not only educates visitors and local residents but also contains an archaeology laboratory to support further historical documentation. Finally, QLF also conducts themed cruises of Labrador and Newfoundland's northern shore, showcasing the region's unique natural and cultural heritage products. For example, in 2001 the organisation led the 'Atlantic Sagas' 14-day expedition cruise, which highlighted the history of occupation of Greenland, Labrador and Newfoundland by indigenous peoples, the Norse and more recent settlers such as the Irish and English. These ecotourism activities raise the tourist's awareness of the region's natural and cultural heritage and increase economic development opportunities within the area (QLF, 2001).

Both QLF and GREMM devote a large portion of their time to research and education in their efforts to achieve the conservation of natural and cultural heritage. They view marine ecotourism as an integral part of this conservation mission. They also understand that marine ecotourism brings both opportunities and problems, and that adherence to the principles of marine ecotourism through the implementation of appropriate management and planning approaches is the key to tapping the potential of marine ecotourism for conservation. Research and education have been essential components of this effort. However, other tools, such as partnerships with local stakeholders, innovative programming and marketing, the use of standards, and so on, have also been essential ingredients in their efforts (see Hull, 2000).

Conclusion

The six case studies presented in this chapter illustrate some of the positive actions NGOs have taken in ensuring that marine ecotourism can be an effective tool for achieving conservation. From these case studies, three specific criteria emerge as requirements for success. The first is an appropriate setting – marine ecotourism businesses and attractions will not be a success if they do not meet fundamental elements associated with viable tourism destinations, such as access to markets and 'world class' products and experiences. While many coastal settings have stunning natural and cultural resources, they may be too isolated from potential ecotourists or may lack adequate services. NGOs can tackle the latter through skills training and product development initiatives, in collaboration with and with the approval of community members. This is entirely contingent upon the second criterion for success found in these case studies, which is adequate resources. These resources include both financial as well as intellectual capital.

A third and final criterion for success is adequate and appropriate planning and management associated with the implementation of marine ecotourism. This includes collaboration and information sharing with communities as equals in the implementation process and 'owners' of the shared coastal resource. It also includes 'buy-in' by resource managers and other related government agents. A final, often neglected, yet critical group of stakeholders in this planning and management process is the tourism industry, particularly its marketing and infrastructure system that includes promotion and distribution tools as well as pricing and product development strategies.

In sum, appropriate planning and management, resources and settings are critical. No marine ecotourism product or destination fostered by NGOs can hope to be a success without these essential components

References

Baleines Endirect (2002) Online documentation, 10 January. [http://www.baleinesendirest.net]. (Also available in English at [http://www.Whales-online.net/indexe.html].)

Bookbinder, M.P., Dinerstein, E., Rijal, A., Cauley, H., and Rajouria, A. (1998) Ecotourism's support of biodiversity conservation. *Conservation Biology* 12 (6), 1399–404.

CSQ (1999) Ecotourism, sustainable development, and cultural survival: Protecting indigenous culture and land through ecotourism. *Cultural Survival Quarterly* 23 (2), 25–59.

Dixon, J.A. and van't Hof, T. (1997) Conservation pays big dividends in Caribbean. *Forum for Applied Research and Public Policy* 12 (1), 43–8.

Eagles, P.F.J. (2000) International trends in park tourism and ecotourism. Paper presented at the Fourth International Conference on the Science and Management Protected Areas Association.

Font, X. and Buckley, R.C. (2001) *Tourism Ecolabelling: Certification and Promotion of Sustainable Management*. Wallingford: CAB International.

Garrod, B., Wilson, J. and Bruce, D. (2001) *Planning for Marine Ecotourism in the EU Atlantic Area: Good Practice Guidance*. Bristol: University of the West of England.

Goodwin, H.J., Kent, I.J., Parker, K.T. and Walpole, M.J. (1997) Tourism, conservation and sustainable development: Vol. 1, Comparative Report. Unpublished report produced for Department for International Development, UK.

Halpenny, E.A. (2002) *Marine Ecotourism: Impacts, International Guidelines and Best Practice Case Studies*. Burlington: The International Ecotourism Society.

Honey, M. (1999) *Ecotourism and Sustainable Development: Who Owns Paradise?* Washington, DC: Island Press.

Hull, J. (2000) *Ecotourism Handbook: Protected Areas and Local Communities: Case Studies from the Middle East and North America*. Ipswich: Quebec Labrador Foundation.

Lindberg, K., Enriquez, J. and Sproule, K. (1996) Ecotourism questioned: Case studies from Belize. *Annals of Tourism Research* 23 (3), 543–62.

Lindberg, K. and Halpenny, E. (2001) Protected area visitor related funding mechanisms: Country review. Draft online document. 10 January [http://www.ecotourism.org/retiesselfr.html].

Marion, J.L. and Farrel, T.A. (1998) Managing ecotourism visitation in protected areas. In K. Lindberg, M. Epler Wood, and D. Engledrum (eds) *Ecotourism: A Guide for Planners and Managers*, Vol. 2 (pp. 155–82). North Bennington: The Ecotourism Society.

Mastny, L. (2001) Travel light: New paths for international tourism. *Worldwatch Paper* 159. Washington DC: Worldwatch Institute.

McLaren, D. (1998) *Rethinking Tourism and Ecotravel: The Paving of Paradise and What You Can Do to Stop It!* West Hartford, CT. Kumarian Press.
Mehta H., Baez, A. and O'Loughlin, P. (eds) (forthcoming) *International Ecolodge Guidelines.* Burlington: The International Ecotourism Society.
Norse, E.A. (ed.) (1993) *Global Marine Biological Diversity: A Strategy for Building Conservation into Decision-making.* Washington, DC: Island Press.
Quebec Labrador Foundation (QLF) (2001) Culture and heritage. *Compass: The QLF Magazine.* 12 (1), 16–22.
Rainforest Alliance (2001) Online document. [http://www.rainforest-alliance.org/programs/sv/certification-standards.html] (Accessed 10 January 2002).
Smith, V. (ed.) (1989) *Hosts and Guests: The Anthropology of Tourism.* Philadelphia: University of Philadelphia Press.
TIES (1995) *Ecotourism Guidelines for Nature Tour Operators.* North Bennington: The International Ecotourism Society.
Tourism Queensland (2000) NEAP II: Australia's Nature and Ecotourism Accreditation Program. Online document. http:\\tq.com.au/qep/neap.htm and http://www.tq.com.au/home.htm (Accessed 10 January 2002).
UNEP (1998) *Ecolabels for the Tourism Industry.* Paris: United Nations Environment Programme.
Walpole, M. J. and Goodwin, H. J. (2000) Local economic impacts of dragon tourism in Indonesia. *Annals of Tourism Research* 27 (3), 559–76.
Weaver, D. (1998) *Ecotourism in the Less Developed World.* Wallingford: CAB International.
WWF (2000) Online document. [http://www.panda.org/endangeredseas/mpa/] (Accessed 10 January 2002).

Chapter 7

Sipadan: An Over-exploited Scuba-diving Paradise? An Analysis of Tourism Impact, Diver Satisfaction and Management Priorities[1]

GHAZALI MUSA

Introduction

The less developed world often has a great variety of unspoiled natural environments, such as beautiful forests, abundant wildlife, high mountains and exotic tropical islands. However, again and again one finds unsustainable tourism management. Developing countries aim largely, if not exclusively, at economic gain. Little consideration is given to conservation strategies or to sustainable tourism management which is directed at positive socio-economic change that does not undermine the ecological and social system upon which communities and society are dependent.

Ecotourism management is currently seen as a sustainable option for marine tourism. In many developing countries, even though conservation awareness among certain organisations is improving, tourism development often becomes a victim of authority interested in immediate economic gain rather than long-term preservation of the environment and concern for local interests. This chapter will examine issues of importance in managing marine ecotourism in Malaysia. An extreme case study, Sipadan, which is grossly mismanaged, is described, this being based on a literature survey and study undertaken by Musa (1998). Management priorities required in order to rehabilitate and safeguard the island are suggested.

Management Issues of Marine Ecotourism in Malaysia

Malaysia is a federation of 13 states. It is blessed with 4800 km of beaches and 1007 offshore islands. Thirty-eight offshore islands in Malaysia have been established as marine parks under the Fishery Act of 1985, with protection covering the area from the low watermark to two nautical miles offshore. The responsibility for the development, administration and management of these marine parks was given to the Department of Fisheries. Hiew and Yaman (1996) identify two management problems: first a lack of conservation awareness, at almost every level; and second a lack of integrated management among tourism stakeholders.

Even though the two nautical miles offshore of the marine park of an island are protected, the islands themselves are under the jurisdiction of individual states. Contradictions in management are often seen. The park authorities (under the federal government) have no power to stop the over-development of the island if the state government permits it.

Coastal resources have various stakeholders, such as federal states, fisheries, private companies, researchers, and many others. Frequently, these stakeholders are operating with different objectives, resulting in conflicts in use that pose serious threats to sustainable management. Environmental issues and concepts of sustainable management are relatively new to this developing country, and developers and policy makers are lacking in general, as well as specialist knowledge in this field. Many decisions pertaining to the use of land are made at the local or state level, and are heavily influenced by immediate economic gain, licit or illicit.

Scientific research is conducted on a largely *ad hoc* basis, with the result that scientific recommendations are often either difficult to access or are ignored by developers and the government. A clear example of environmental recommendations being ignored is the development of Redang Island. A thorough impact assessment study predicted that the major resort development proposed for the island would lower the freshwater aquifer, cause erosion to unstable slopes and inflict serious harm to the reef encircling the island. Despite this prediction, the state government sanctioned a private company to proceed with the project. Following the resort's construction, the slopes started to erode, resulting in siltation, which subsequently eliminated many coral reef species and clouded the water to the point where the remaining species became hard to see (Manning & Dougherty, 1995).

Wong (1996) argued that Malaysia's offshore islands are being promoted internationally before the necessary laws, management plans, facilities and manpower are in place. Laws are out-dated and have little concern for proper environmental regulation (Hiew & Yaman, 1996).

Sipadan: The Victim of the Worst Tourism Mismanagement in Malaysia

Sipadan is a disputed island, claimed by three countries, Malaysia, Indonesia and the Philippines, but it is currently administered by Malaysia. It is situated in North Eastern Borneo (5.5 degrees North and 120 degrees East, see Figure 7.1). The size of the island is only 16.4 hectares.

Geologically, the island was formed by a volcanic thrust from the Celebes Sea floor 600m below the surface, forming a mushroom-like island which rises only 3m from low tide water mark (Figure 7.2). The unique underwater wall of the island provides glorious underwater landscapes, clad in many genera of soft and hard corals. Over 3000 fish species are recorded in its waters, a diversity similar to Australia's Great Barrier Reef (Jackson, 1997). Sharks and turtles can be seen on every dive. The World Wildlife Fund states that 'no other spot on the planet has more marine life than this island' (http://www.deepdiscoveries.com/sipadan.htm). In the island forest there are 65 species of plants and 12 species of birds (UKM, 1990).

Historically, the first official document relating to the island was issued by the government of British North Borneo on 6 May 1916. It gave two persons the right to collect turtle eggs in Sipadan. In 1933, the island was designated as a bird sanctuary for the protection of an underground nesting bird (the Nicobar pigeon). The image of Sipadan as an underwater paradise emerged dramatically following a documentary with worldwide circulation produced by the famous underwater explorer, the late Jacques Yves Cousteau, in 1988. He described Sipadan as 'an untouched piece of art' and 'one of the last jewels on the planet' (cited in Malmstrom, 1997: 13). This publicity was also the beginning of what could lead to irreparable damage.

Tourism development on the island to date is staggering, and threatens its small and fragile resources. Malmstrom (1997: 13) commented that 'today, nature's

Figure 7.1 The location of Sipadan

Figure 7.2 A bird's eye view of Sipadan and its reef (*Source*: Borneo Divers and Sea Sports)

finished masterpiece has become one of mankind's "work in progress", and the original "piece of art" is now in danger of being ruined'.

Despite the growing concern on the possible impact of tourism on the island (Chua & Mattias, 1978; Malmstrom, 1997; Mortimer, 1991; UKM, 1990; Wood, 1981; Wood *et al.*, 1993, 1995), by 1998 there were six diving operators on the island. They collectively accommodated 360 divers per day and were run by approximately 180 staff (Musa, 1998). One of the resort managers commented, 'pumping water from the small well for the use of 360 people (actually 540 in total when staff are included) – in addition to pumping out their waste – means the island is virtually caving in on itself' (cited in Malmstrom, 1997: 13). Divers also stay at a nearby island (Mabul) where there are three dive operators. According to a dive master, during peak seasons 500 divers with an average of three dives each per day could be swarming around the 13 dive sites along the underwater wall of the tiny island. Of the island's land area 50% has been developed. The clearing of the vegetation and the lighting produced by the six resorts has had a detrimental impact on the turtles. Crowding underwater is common, as no fixed arrangements have been made by dive centres as to where they should be taking their divers at any particular time. In the evening, two-thirds of the beachfront is covered by moored dive boats, seriously affecting the ambience and tranquility of the island.

Scientific Research on Sipadan

Several scientific studies have been conducted on Sipadan (Chua & Mattias, 1978; Mortimer, 1991; UKM, 1990; Wood, 1981; Wood *et al.*, 1993, 1995). All of them, without exception, propose that the island should be designated as a marine park.

The first scientific research was conducted on the area by Chua and Mattias (1978), who were investigating the impact of oil spillage and accidentally discovered the island's unique qualities. They were the first to propose its designation as a marine park. Wood (1981) conducted an extensive marine survey and was dazzled by the underwater treasures the island possessed. She made a proposal similar to Chua and Mattias on the basis of the island's unique underwater geological formations, outstanding marine scenery, interesting flora and fauna, and the importance of the island for the breeding of turtles. Wood also suggested that in view of the tiny size of the island and the fragility of its ecosystem, no overnight stays should be allowed. Furthermore, Wood recommended that the island should only be available for education, research and appreciation. No action was, however, taken by any authority on these recommendations.

When a research team from the National University of Malaysia (UKM) arrived in 1990 to assess the impact of tourism development on the island, 8% of the island had been developed by three establishments: two dive operators and the wildlife department. Among the problems identified by the research team were coral damage, an inappropriate sewage system, the clearing of vegetation, increased erosion, sedimentation and plants and humans competing for limited underground water sources. Curiously, even though the island is so small, the research team recommended 'zoning' for recreational users on the island. They also proposed that the island's wonders should be 'appreciated' and said that the island could easily afford 40 overnight tourists.

The clearing of vegetation, excessive lighting and over-development for tourism have led to serious concern for the turtles on the island (Mortimer, 1991). The number of turtles nesting on the beach has been reduced by 31.7% from 1470 turtles in 1992 to 1001 turtles in 1997 (SWD, pers. comm.). Mortimer conducted a survey on turtle nesting in Sipadan and suggested that turtles should be allowed to incubate their eggs in their own natural nests, which are displaced continually by the development of accommodation for the tourists. Natural nesting is also important, to ensure the gender balance of turtles (UKM, 1990). The UKM team observed that green turtles prefer undisturbed beaches and being under low overhanging branches along the coastline for nesting. Cool nests (temperature less than 28 degrees in temperature) predominantly produce male turtles, while warm nests (above 30 degrees temperature) produce predominantly female turtles. This is contrary to the present management regime, whereby turtle eggs are removed from their natural nests every morning by the park rangers and incubated in pre-prepared nests, which are situated in an area exposed to the sun. While this prevents the eggs from being damaged by predators (the area is enclosed by fences and nets), it is doubtful whether such a management regime can ensure the gender balance necessary for the survival of the species. The northern rim of the island, which is the easiest access for turtles to lay eggs on the beach, is very crowded with tourist accommodation and moored dive boats. A dive resort has even built a concrete wall, in order to prevent erosion to the beach fronting their establishment.

Wood and her associates returned to Sipadan in 1993 and saw that 13.5% of the island had been developed for three dive operators. She reported that coral reefs were healthy, with an abundance of fish. Contrary to her earlier proposal in 1981, Wood *et al.* (1993) proposed that the island should be able to accommodate 50 overnight tourists. She again proposed that the island should be designated as a marine park. Wood and her associates visited the island again in 1995 and expressed grave concern about the over-development of the island, which had by then extended to 33.4% of the land area. She regretted that political and economic considerations seemed to dictate events on Sipadan. Even though Davis (1993) states that scuba-diving activities are less likely to cause damage to marine life in comparison to natural causes, over-development of the land may indeed do so. The example of Redang was mentioned earlier, where the development of a resort led to erosion and siltation, subsequently reducing underwater visibility and eliminating many coral species (Manning & Dougherty, 1995).

In an interview, Holland (pers. comm.) stated that 'over the last 18 years I have watched one of the world's greatest diving spots and a Malaysian heritage, slowly being destroyed by greedy operators who didn't have a care about the long term future of the island. They just cared about the short-term quick-cash turnover'. Holland expressed his worry that if the trend carries on in the present way, the future of Sipadan will be very bleak. He stated that the government should have not allowed so many companies to become established on the island: 'lack of overall and effective development planning and control and eager entrepreneurs are the main reasons for what has happened and is happening to the ecosystem of Sipadan'. Holland, who literally found the island in the early 1970s with his group of fellow divers, continued to explain how lack of planning control was the main reason for the island's deterioration. He stated that following the study undertaken by the World Wildlife Fund in 1989 it was recommended that only one dive operator should be allowed to operate on the island. Soon after the establishment of Borneo Divers (the first dive operator) on the island, an influential businessman received approval to set up another dive operation alongside this. Soon after that, another dive operation was given permission. Holland added, painfully: 'there was no restriction on how many cottages they could build: in fact there were no restrictions at all'.

An Investigation of Tourists' Satisfaction and the Impact of Tourism Development

Methodology

Following the news of haphazard tourism development in Sipadan, and the suggestion by Wood *et al.* (1995) that a tourist satisfaction study should be carried out,

Musa (1998) conducted the study with three main objectives: first, to identify the diver profile, second, to examine overall diver satisfaction and the determinants of their satisfaction and third, to ascertain the impact of tourism development on the island.

Data collection was made by means of a self-completed questionnaire. Due to the controversial nature and the difficulty of measuring satisfaction with the SERVQUAL model introduced by Parasuraman *et al.* (1988), the study employed the Nordic approach of SERPERF, as proposed by Gronroos (1990). Gronroos measured satisfaction based solely on actual quality performance and its perception by consumers.

Overall satisfaction was measured using a nine-point Likert scale. All individual items of the tourist's experience (activities, services, facilities, nature) were measured separately, following the approach of Danaher and Aweiler (1996) and Gnoth (1990). This approach was taken because tourists' motivations are complex and tourists are looking for satisfaction across a range of aspects of the tourism product. As Gnoth (1990) argues, overall satisfaction measures only the overall emotional response to the entire experience and is inadequate for diagnostic purposes.

In the questionnaire, 29 variables relating to scuba-diving satisfaction were identified, covering aspects such as natural beauty, activities, services and facilities. The variables were derived from scuba-diving studies conducted by Tabata (1992), Lim and Spring (1995) and Davis *et al.* (1996), and pre-testing and pilot studies were conducted among divers in New Zealand and Malaysia. Divers were asked to rate their response to a number of statements along the continuum of a five-point Likert scale, ranging either from 'very good' to 'very poor' or from 'strongly agree' to 'strongly disagree'.

Since Ryan (1995) stated that positively-skewed responses are often seen in overall satisfaction studies, in this study open-ended questions in the form of best aspects and worst aspects, as well as a comments section, were provided in order to elicit a broader picture of diver satisfaction.

The sampling method involved the personal delivery of the questionnaire to all divers on the island, on the morning of their departure or the day before in order to minimise interference with their activities on the island and thereby increase the response rate. The survey period extended for 17 days in November 1997. A total of 314 questionnaires were collected for final analysis and the response rate was calculated to be 93%. The high response rate was due to the questionnaire administration being done in a captive location and the level of concern evidenced by divers about the future of the island paradise.

Results and discussion

Twenty-two different nationalities were sampled in this study, of which the top seven were Japanese (27.7%), British (13.2%), Malaysian (12.6%), Korean (8.7%) and Swiss (6.5%), followed by German and American (4.8% each). Males constituted 64.8% of the divers while 35.2% were females. The gender distribution was similar to the surveys conducted by Skin Diver magazine and Underwater USA (see Tabata, 1992). The average age for divers at Sipadan was 34.9 years, which is slightly lower than the figure given by Tabata (1992). Seventy-one per cent of divers had higher qualifications than high school. The majority of respondents in Sipadan (69%) were experienced divers (holding a minimum of Advance Water Certificate), while 31% were novices, holding only an Open Water Certificate. Most divers were 'hardcore', as categorised by Rice (1987), as they came to Malaysia solely for scuba diving (79.9%). The average length of stay was 6.6 days, which is slightly higher than the 5.4 days of the average stay of international tourists in Malaysia (ASTR, 1997).

The vast majority of the divers (97.8%) rated their experiences as 'satisfied' to 'extremely satisfied' (mean = 2.01). Only seven divers (2.2%) rated their experience as 'neutral' to 'extremely dissatisfied'. The dissatisfaction, as expressed in the comment section, was mainly attributed to poor underwater visibility, crowding and over-development on the island. Fifty–nine per cent of the experienced divers (who had dived in more than five countries) rated Sipadan as the world's best dive-site. Similarly to the satisfaction studies conducted by Ryan (1995) and Rubenstein (1980), in this study females were likely to achieve higher overall satisfaction than males ($\chi^2 = 23.807$, $df = 3$, $p = 0.000$). No significant correlation was found between satisfaction and any of the following variables: educational achievement, nationality, number of countries that divers have dived in before, diving certification and age.

Similarly to the overall satisfaction level, divers rated most individual satisfaction items positively. To help the discussion and possible management plan, the rated items were grouped into three categories. First, attributes with a mean score of less than two were considered to be central determinants of diver satisfaction, which should therefore be maintained and enhanced. Second, attributes with a mean between two and three were considered to be areas that may have some room for improvement. Third, attributes with a mean of more than three were seen as requiring urgent action. Table 7.2 shows the determinants of diver satisfaction in Sipadan.

Abundance and variety of marine life (mean = 1.35) were considered to be the greatest assets of Sipadan. General staff (mean = 1.45) and the services of dive masters (mean = 1.67) were also highly rated in terms of satisfaction by divers. A unique feature of Sipadan is the easy access to the dive sites (mean = 1.55), as all diving sites are less than five minutes away by speedboat. This finding is confirmed further by the answers given in the open-ended questions: 'marine life', 'easy access and unlimited dives' and 'marine environment' comprised 68.1% of respondents' best aspects of Sipadan diving experiences. Social aspects, such as good diving buddies (mean = 1.47), and meeting people and making friends either with staff or other divers (mean = 1.91), were also significant contributions to satisfaction with the experiences offered on the island. These were also cited as being among the best aspects of divers' visits in the open-ended questions.

The second group of diving attributes, for which the ratings suggested some

Table 7.2 Determinants of diver satisfaction in Sipadan

Scuba-diving attributes	*Frequency*					*N*	*Mean*
	VG	*G*	*N*	*P*	*VP*		
1. Marine life	223	70	19	0	0	312	1.35
2. Friendly/helpful staff	196	90	21	1	1	309	1.45
3. Good buddies	204	67	29	4	2	306	1.47
4. Water temperature	192	88	28	1	0	309	1.48
5. Easy dive access	191	72	35	7	1	306	1.55
6. Professional and efficient dive master	165	98	33	9	4	309	1.67
7. Efficient staff	153	105	45	6	1	310	1.70
8. Underwater geological formation	117	124	61	4	2	308	1.86
9. Making friends	142	85	55	12	12	306	1.91
10. Coral reef	124	110	57	18	2	311	1.92

VG = Very good; G = good; N = Neutral; P = Poor; VP = Very poor.

improvement may be required, were accommodation (mean = 2.62), ground transportation (mean = 2.55), rental equipment (mean = 2.34), 'tourist information' (mean = 2.56) and 'interpretative facilities' (mean = 2.48). The third group of attributes which requires urgent addressing comprised crowding (mean = 3.53), which will be discussed in the following section.

Tourism impact

Sustainable tourism development in Sipadan faces poor chances of survival because of the impact of the developments that have already been permitted. Over 90% of the negative comments were related to crowding and over-development. Disappointment with the way in which the island was being run were illustrated vividly in the comments section, including comments such as 'the island is over developed at the risk of ruin and 'the island is one of the prime examples of loving the place to death'.

The impacts of tourism development in Sipadan as perceived by divers can be grouped into two categories: physical impacts, which include 'underwater visibility' (mean = 2.67) and 'underwater pollution' (mean = 2.39), and social impacts, which include 'crowding' (mean = 3.53) and 'noise' (mean = 2.41).

Crowding was evidently the most worrying social impact as perceived by divers (mean = 3.53). Crowding in Sipadan is a phenomenon that can not only be seen on the beach but also underwater. Underwater, in the midst of bubbles produced by themselves, divers often get confused and follow other groups while diving. This may lead to panic on the part of the dive masters, who may fear that one of the members of their group has disappeared. No significant correlation was found between the perception of crowding and either nationality or satisfaction. Two New Zealand divers rated the island as extremely crowded, with a mean of

1.00 but at the same time they were extremely satisfied with their entire experience (mean = 1.00). This observation could be attributed to 'coping strategy' whereby, in the case of obstructed recreational desires, users modify their perception, behaviour and recreational priorities in order to gain satisfaction (Higham, 1996; Stankey & McCool, 1994).

Noise (mean = 2.41) was considered the second most serious social impact on Sipadan. There is no part of the island from which generator noise cannot be heard. Apart from generators, noise is made by divers, loud music from radio and TV, and even the staff themselves. Noise not only disturbs wildlife but it is also undesirable to divers who seek peace and tranquility on the island.

Even though underwater visibility is often seen as the second most important determinant of diver satisfaction after marine life (Davis *et al.*, 1996; Lim & Spring, 1995; Tabata, 1992; Wong, 1996), in Sipadan the attribute does not appear within the top ten of diving attributes (mean = 2.67). Poor visibility could be caused by rain, tidal changes or plankton. However, in Sipadan it may be a warning sign of more serious environmental consequences of over-development. Leakage from sewage disposal and the use of fertilisers (for flowers and shrubs) could contaminate the seawater and subsequently lead to eutrophication (excessive nutrients in the water). Septic tanks, which are widely used in Sipadan, provide no guarantee against leakage of nutrients and pathogens into the sea (Lim, 1997). Furthermore, flushing the toilet with seawater, which is a common practice in Sipadan, reduces effective biochemical degradation. Cross-tabulation shows a significant correlation between visibility and 'unmet' expectations ($\chi^2 = 8.208$; $df = 2$; and $p = 0.017$). If visibility worsens, visitor satisfaction and therefore probably the sustainability of the dive operations in Sipadan will be compromised.

There was some degree of concern regarding underwater pollution (mean = 2.39). Divers reported cans, plastic bags and traces of oil in the water. These not only reduce the amenity of the island but they also endanger marine life. Plastic bags have been known to choke to death the turtles that mistake them for jellyfish (which form the turtles' diet). Often speedboats move at great speed around the island, which not only endangers divers, but also turtles. In fact, the sight of turtles with apparent propeller injuries disturbed some divers. Turtles – which need to breathe at least once every hour – are highly susceptible to these injuries, due to their sluggish movement on the surface while breathing.

In the open-ended question, 11.5% of the worst aspects of Sipadan were concerned with the physical impact on the land itself. These included rubbish on the beach as well as the forest behind the resort having become a rubbish dump. One diver commented 'I am shocked at how polluted the island is with litter, washing lines, generator noise, sewage smells and boat fumes'. Musa (1998) also observed broken drains and sewage pipes behind the resort, which not only cause unpleasant smells but also endanger divers by possible food contamination. Some tourists are also concerned at the sound of dynamite fishing – practised by local fisherman from nearby islands – which could be heard by divers even though the site is miles away. Evidence of old coral damage could also be seen.

Recent Developments

Despite all the proposals that Sipadan should be designated as a marine park, the authorities remain silent. Musa (1998) suggests that the Malaysian government may be reluctant to take action by the way of control because Sipadan has a disputed status. Such action might be construed as pre-empting the decision of the International Court of Justice, to which the dispute has been referred. Close to a decade after the referral, the matter is still with the court. According to Bajerai (pers. comm.), the manager of Borneo Divers (the main diving operator on the island), several developments have occurred in Sipadan. The Federal Security Council has limited the numbers of overnight divers to 80. Only 20 divers staying elsewhere are allowed to dive each day, which means the overall maximum number of divers allowed to dive in one day is 100. The number of staff allowed on the island is 35. Although this limitation on numbers is not based on any scientific study, the decision could be the first step in slowing down the exploitation and moving towards the conservation of this unique marine area.

A 'Resort Consortium' was formed in 1998. Holland (pers. comm.) states that the consortium was set up in the hope that all diving companies on the island could get together and build just one resort on the island and demolish all the other resorts. However, the decision is still pending. Bajerai (pers. comm.) states that 'much disagreement occurs among resort owners, mainly over their right to take more customers'.

On 23 April 2000, Sipadan created another mark on the world map as a target of terrorism (http://www.e-borneo.com/news/smain-a.html). Twenty-one divers were kidnapped from the island by Abu Sayyaf extremist rebels from the southern Philippines, for political and monetary gain. The governments of Japan (Japan being Sipadan's biggest market) and the US subsequently issued warnings to their citizens not to take their holiday in Sipadan and disclaimed responsibility for their safety should such warnings be ignored. Following the incident in 2000, tourist numbers in Sipadan dropped by 60–100% among diving operators in the subsequent six months. Abu Sayyaf may unintentionally have reduced the rate of over-exploitation of the island, and momentarily given a breathing space to the marine life of Sipadan. Meanwhile the Malaysian government has taken rapid action to protect tourists. Sipadan and the islands around it are now heavily guarded by the military. Consequently, Sipadan is said to be a safe place to visit and tourist numbers are now increasing again. This may allow the destruction of the island to continue once more. An experienced journalist, familiar with Sipadan and its history, says the detrimental developments on the island 'are founded purely on greed' (Anon, pers. comm.).

Management Priorities

Today, ecotourism is seen as a sustainable option for marine tourism. Ecotourism means tourism development through activities that are based on nature, ecologically sustainable, environmentally educative, locally beneficial and generating visitor satisfaction (Dowling, 1997). It has a strong emphasis on eco-friendly management strategies and energy and waste minimisation. Shurcliff and William (1991) argue that successful management is dependent on a good understanding of the nature of environmental impact, agreeing on management objectives and strategies, determining what action will achieve the outcome and who can best undertake this action, and then monitoring and evaluating it.

Judging from the impact observed on the island, and evidence of disturbed ecological balances as a result of over-development of the island, Sipadan tourism development is not sustainable and cannot be considered to be compliant with ecotourism management. Holland (pers. comm.), a veteran manager in one of the resorts on the island states:

> There is no way the tourism development on Sipadan can be considered as ecotourism. It is just the opposite. Even though the marine life has surprisingly increased from when I first came to the island, the opposite is happening to the island itself: massive erosion of the north beach, caused by operators building concrete walls or placing sandbags to stop erosion around their resorts, the pumping of well water that has depleted the freshwater table to the point that all the existing ground water is now contaminated by salt water, and even sewage is not properly treated.

Even though tourist satisfaction for the holiday is high, the main contributing factor is the marine life and corals of the island. Interpretative facilities on the island are limited, and in the comment sections of the questionnaires, none of tourists mentioned that they had learned anything about marine life or its conservation. In fact, the majority of them viewed Sipadan tourism development with sadness and anger. At the time of the survey in 1998, some dive operators used the services of foreigners as dive masters. While this may be prompted by the need to ensure that divers who do not speak English well (especially the Japanese) feel at home, the use of foreigners does not correspond to ecotourism principles, which state that ecotourism should benefit local people. Arguably, ecotourism operators should employ only local people in order to support the local economy.

The study by Musa (1998) suggests several management priorities. He proposed again the idea suggested by Wood (1981) that due to the small size of Sipadan and its susceptibility to tourism impact, no development should have taken place on the island at all. All the resorts on the island should be relocated to nearby islands or the mainland. The size of Sipadan makes sustainable tourism development (i.e. development without disrupting the ecological balance on the island) virtually impossible. A Malaysian diver commented: 'leave the island alone, all dive operators should not be on the island. It may not be a money-spinner but it is a "land bank" heritage for future generations'. The damage done by the development on the island requires urgent attention by both responsible local and international bodies, and rehabilitation needs to be carried out on the island to re-establish the delicate ecological balance. Marketing should place more emphasis on diving quality rather than the convenience of divers, which is the emphasis of the current marketing undertaken by operators.

When asked what he feels about Sipadan's future, Holland (pers. comm.) stated:

> As an optimist, I still feel that there is a chance for Sipadan. Something must be done immediately. There is still that small chance that things can be changed around if the operators are forced into implementing the consortium, as it supposed to be: just one resort, maximum 80 guests; to totally ban all other divers from diving Sipadan; the rest of the resorts to be completely demolished and the island cleaned of all the garbage that has been thrown around for years; finally to hand over the running and operating of the resort to professionals of proven track record, a focus on conservation and the strict observance of all rules and regulations. For me this is Sipadan's last and only chance to survive.

Furthermore, Sipadan should, without any further delay, be designated as a marine park. The management should be integrated among the various stakeholders. Government policy must place a greater emphasis on the protection of the island's ecological balance as opposed to immediate economic gain. An ultimate environmental threshold (UET) study should be conducted to determine the sensitivity of the geomorphological structure, vegetation and fauna in relation to existing developments. Two other studies urgently needed are the determination of the 'carrying capacity' or the 'limits of acceptable change' (LAC) and a proper 'environmental impact assessment' (EIA). Along with these, the appropriate authority should draw up a strategic plan, with limits and rules governing the island, as well as an appropriate system of enforcement.

In view of its small size, zoning for various users as suggested by UKM (1990) is not practical. The sole recreation permitted should be scuba diving. The authorities should review the use of the island for snorkellers, picnickers and preliminary or basic scuba-diving certifications such as the 'Scuba Discovery' or the 'Open Water Certificate' of persons who are unaware of or inexperienced in the limits of behaviour required by the very fragile environment. The operators should adopt ecotourism management techniques, which emphasise tourism development and activities that are ecologically sustainable, environmentally educative, locally beneficial and generate visitor satisfaction. Eco-friendly management strategies such as energy, water and waste minimisation and the use of recycling should be introduced. Divers should be educated to appreciate the importance of complying with these guidelines.

Cooperation between operators is also needed in order to minimise social, physical and ecological impacts on the island. Centralised energy, water, garbage disposal and safe sewage treatment should be introduced. Furthermore, centralised transport services would minimise current problems relating to the independent transportation of divers. The noise from generators and compressors could be reduced by placing them in soundproof compartments. Regular and extensive litter picking of the beach and the interior of the island is also required.

A speed limit should be enforced on boats in proximity to the island. This is not only to prevent injuries to marine life such as turtles but is also important for divers' safety and ensuring the tranquillity of the island.

Musa (1998) suggested that leaving Sipadan's future only to Malaysia, Indonesia or the Philippines, all of whom have so far been unable to give very much attention to environmental conservation, may not be adequate to safeguard the island. The sustainable management of Sipadan requires international support, including support from divers. The immediate need is for its designation as a marine park but steps could be taken to obtain the designation of the island as a World Heritage Site under the Convention Concerning the Protection of the World Cultural and Natural Heritage (the 'World Heritage Convention', 1972), to which all three countries are signatories. This would ensure that the island's management would be monitored in the international interest by the International Union for Conservation of Nature and Natural Resources (IUCN), under UNESCO's World Heritage Trust.

Conclusion

Sipadan Island provides a very rare opportunity for the study and enjoyment of the tropical underwater world. Within two decades, the island has emerged from a virtually unknown turtle nest to a scuba-diving Mecca, a subject of international dispute and a terrorist target. Scientific studies and media coverage continue to express concern over the exploitation of the island but the suggestion that the island should be designated as a protected marine park has fallen on deaf ears. However, the island's treasures are continuing to be exploited at a ruinous rate. Human greed has taken precedence over turtle breeding sites, freshwater for the flora and fauna of the island, underwater visibility and the peace and tranquillity, for which most tourists are looking. Thus, tourism development in Sipadan is seen as unsustainable and remote from the notion of ecotourism management. Divers are highly satisfied with the diving experience in Sipadan and this is mainly due to the richness of marine life, easy access to diving sites and staff efficiency and friendliness. However, they are alarmed at the impact of tourism development resulting in crowding above water and underwater, noise, litter and the poor underwater visibility. It is a hope of this research to capture local and international interest to save the island. Tourism resources in less developed countries are often the victim of a desire for immediate economic gain, inefficient tourism management and political indifference. The late Jacques Cousteau described Sipadan as 'the last jewel of the planet'. If it is to survive as such, effective action is needed and it is needed urgently.

Acknowledgements

The author acknowledges the assistance of Dr James Higham for supervising the research, Dr Peter Mooney for editorial suggestions, all the dive operators in Sipadan for allowing their divers and staff to be interviewed, Borneo Divers, Sipadan Lodge and Bajau for providing logistical support, Ron Holland and Agill

Bajerai for making themselves available for interview, and last but not least all the divers who completed the questionnaires.

Note

1. This chapter was originally published in *Tourism Geography* 4 (2), 195–209 (2002), and appears with the permission of Taylor & Francis Ltd (http://www.tandf.co.uk/journals).

References

Annual Statistical Report (1997) Annual tourism and statistical report, Malaysia Tourism & Promotion Board.

Chua T. and Mattias R. (1978) Sabah coastal resource. Unpublished survey by University of Science Malaysia.

Danaher, P.J. and Arweiler, N. (1996) Customer satisfaction in tourist industry: A case study of visitors to New Zealand. *Journal of Travel Research* 34 (Summer), 89–93.

Davis, D. (1993) Scuba-diving: Conflicts in marine protected areas. *Australian Journal of Leisure and Recreation* 3 (4), 67–72.

Davis, D., Banks, S.A. and Davey, G. (1996) Aspects of recreational Scuba-diving in Australia. In G. Prosser (ed.) *Tourism and Hospitality Research: Australian and International Perspectives* (pp. 455–66). Canberra: Bureau of Tourism Research.

Deep Discoveries. Online Documentation. 1 September 2001 http://www.deepdiscoveries.com/sipadan.htm.

E-Borneo. Online documentation. (4 September 2001) http://www.e-borneo.com/news/smain-a.html.

Dowling, R.K. (1997) Plan for the development of regional ecotourism: Theory and practice. In C.M. Hall, J. Jenkins and G. Kearsley (eds) *Tourism Planning in Australia and New Zealand: Cases, Issues and Practice*. Sydney: Irwin Publishers.

Gnoth, J. (1990) Expectation and satisfaction in tourism: An exploratory study into measuring satisfaction. Unpublished PhD thesis, University of Otago.

Gronroos, C. (1990) *Service Management and Marketing: Management in the Moment of Truth in Service Competition*. Lexington, MA: Lexington Books.

Hiew, K.W.P. and Yaman, A.R.G. (1996) The Marine Park of Malaysia: Objectives, current issues and initiatives (draft). Paper presented at the workshop on impact measurement of marine parks, 13–14 August, Kuala Lumpar.

Higham, J.E.S. (1996) Wilderness perceptions of international visitors to New Zealand. Unpublished PhD thesis, University of Otago.

Jackson, J. (1997) *The World's Best Dive-sites.* Singapore: New Holland.

Lim, L.C. (1997) Carrying capacity assessment of Pulau Payar Marine Park. Report produced under project No. MYS 3411/96, WWF Malaysia.

Lim, L.C. and Spring, N. (1995) The concepts and analysis of carrying capacity: A management tool for effective planning, Part 3, case study: Pulau Tioman. Report produced under project MY0058. Kuala Lumpur: WWF Malaysia.

Loundsbury, J.W. and Hoopes, L.L. (1985) An investigation of factors associated with vacation satisfaction. *Journal of Leisure Research* 17 (1), 1–13.

Malmstrom, K. (1997) Sipadan under siege. *Action Asia Magazine* 6 (3), 13–15.

Manning, E.W. and Dougherty, T.D. (1995) Sustainable tourism: Preserving the golden goose. *Cornell Hotel and Restauran Adminisration Quarterly* (Apr), 29–42.

Mortimer, A.J. (1991) Recommendations for the management of the marine turtle population and Pulau Sipadan. Report by WWF, Malaysia.

Musa, G. (1998) Sipadan: A survey of the geographical aspects of divers' satisfaction on

Sipadan island. Unpublished dissertation submitted for the Diploma in Tourism, University of Otago.

Parasuraman, A. Zeithaml, V.A. and Berry, L.L. (1988) SERVQUAL: A multiple item scale for measuring consumer perception and service quality. *Journal of Retailing* 64 (1), 12–40.

Rice, K. (1997) Special report: Scuba-diving: Dive market requires specialised skill and information. *Tour and Travel News*, 9 February, 24–7.

Rubenstein, C. (1980) Vacation: Expectations, satisfactions, frustrations and fantasies. *Psychology Today* 14, 62–6 and 71–6.

Ryan, C. (1995) *Researching Tourist Satisfaction: Issues, Concepts and Problems.* New York: Routledge.

Shurcliff, K. and William, A. (1991) Managing ecotouriusm in the Great Barrier Reef Marine Park: Can we manage it together? In B. Weiler (ed.) *Ecotourism Incoprorating the Global Classroom* (pp. 61–4). Canberra: Bureau of Tourism Research.

Stankey, G.H. and McCool, S.F. (1984) Carrying capacity in recreational settings: Evolution, appraisal and application. *Leisure Sciences* 6 (4), 453–73.

Tabata, R.S. (1992) Scuba-diving holidays. In B. Weiler and M. Hall (eds) *Special Interest Tourism* (pp. 141–66). London: Belhaven Press.

UKM (1990) Assessment of development impacts on Pulau Sipadan. Universiti Kebangsaan Malaysia, Kampus Sabah, Kota Kinabalu, Malaysia.

Wong, G. (1996) The need for economic valuation of marine and coastal tourism benefit in Malaysia. Report produced under project No. MY 0062, WWF Malaysia.

Wood, E. (1981) Semporna Marine Park survey: Expedition report and recommendations. Report produced under WWF project MAL 134, Malaysia.

Wood, E., George, D., Dipper, F., Lane, D. and Wood, C. (1993) Pulau Sipadan: Meeting the challenge of conservation. Report produced under WWF project MYS 233/92, Malaysia.

Wood, E., Wood, C., George, D., Dipper, F. and Lane, D. (1995) Pulau Sipadan: Survey and monitoring. Report produced under WWF project MYS 319/95, Malaysia.

Chapter 8
Marine Ecotourism through Education: A Case Study of Divers in the British Virgin Islands

CLAUDIA TOWNSEND

Marine tourism brings with it risks and opportunities, which both public and private sectors have a responsibility and interest in managing effectively. This chapter will concentrate on dive tourism and consider those risks and opportunities, in the context of environmental education and its management benefits. The case study outlined later describes a successful experiment that aimed to reduce direct impact on a coral reef by scuba divers through education.

Marine Protected Areas (MPAs) are frequently designated as 'no use' or limited use zones. Just as has been the case in terrestrial parks, tourism, or 'ecotourism', is often proposed as the 'no-smoke', 'win/win' solution which preserves the area's fragile environment and simultaneously provides alternative or additional livelihoods to local people who may no longer have free access to the natural resources within that protected area.

The experience of terrestrial protected areas has shown that this ideal win/win situation requires specialised and effective management to avoid conflict and maximise potential benefits to local communities and their environment.

If tourism is to be considered a valid option for the sustainable use of MPAs, it must be careful to achieve an optional balance between the negative environmental and social impacts that might result and the potential benefits. It must, in other words, be truly sustainable.

This chapter will look at the potential for education as a tool for MPA managers. The maintenance or improvement of ecological integrity is generally the first priority of managers and purely from that perspective it becomes important that the well-documented, destructive impacts of tourists be reduced and contained. Beyond environmental guardianship, managers generally also have financial responsibilities and a public relations role to play. Well-planned education can assist with all of these jobs.

Scuba diving – Management Threat or Opportunity?

Diver impact studies from coral reefs around the world have not always agreed on the ecological significance of damage caused by scuba divers kicking or banging reef organisms. Both the short- and long-term effects of this damage have been found in recent studies to be more significant than previously thought. The impact is clearly localised to areas used as dive sites, and studies have found it to be further localised to the small area surrounding the mooring or anchor where divers begin and end dives. When seen from a narrow ecological perspective, it is easy to relegate diver impact to a low position on the list of worldwide threats to reefs and other marine habitats. Indeed, when compared with dynamite fishing, various types of pollution and global warming to name but a few, this cause of degradation may well be biologically insignificant.

For a number of reasons, however, it is worth addressing the impact that reef visitors (divers, snorkellers, snuba divers, glass-bottomed boat visitors and so on) can have on the reefs that attract them.

Technology is allowing more and more people to visit coral reefs. The invention of scuba in the 1960s has created an ever-growing number of people willing and able to travel further and further, to experience reefs and other marine environments. These recreational divers are likely to have an exponential impact with regard to the numbers of people diving and the number of sites they visit. In recent years, however, a number of new technologies such as snuba (a cross between scuba and snorkelling) have been developed, which allow people to breathe underwater and reach reefs without the time and training necessary to become a certified scuba diver. The new practice of 'reef walking' has similar implications. Dive certification agencies have competed with initiatives to allow greater numbers of people to experience scuba through the relaxation of age restrictions and the promotion of try-dive experiences (which allow people to dive on a reef with only a couple of hours of training under the supervision of an instructor). This increased access to the underwater environment means that certified divers as a group are only the tip of the iceberg of reef visitors. Certified divers, however, are the only group whose numbers are recorded and therefore the only statistic that can give us an idea of numbers of people visiting reefs under the water. PADI (the Professional Association of Dive Instructors) is the largest diver certification agency but, by their own estimate, they certify only 55% of divers worldwide. They have certified over 10 million divers in the organisation's history (PADI, 2001). The British Sub-Aqua Club (BSAC), the main British diver organisation, has made an estimate based on the limited information available that there are around 6 million active divers worldwide (BSAC, 2001). It is impossible to know the numbers of reef visitors who potentially contact and damage reefs but it is indisputable that age, health, mobility and training are no longer the restrictive factors they once were for access to coral reefs. The only major restriction is the ability to pay for a holiday to a reef area, so the population of tourists being introduced to this underwater habitat is growing year on year. There are risks associated with increased accessibility and visitation but also some very significant opportunities. Education is one way to address the risks and harness the opportunities.

The lack of training necessary for these new forms of reef visitation may mean that reef walkers, snorkellers and others are less aware and less skilled on the reef and may cause more damage to it than trained scuba divers. Evidence of the comparative impacts of these activities is notably lacking but there is unarguably less time available on a trip lasting a few hours for operators to instil a sense of environmental responsibility and environmentally relevant skills than there is over the course of a four-day basic training scuba course.

As reefs become more and more accessible and tourists become more aware of their attractions through books, television programmes, travel marketing and word of mouth, it is inevitable that negative impacts will result if tourism is not managed effectively. What is needed, then, are ways of reaching these tourists with effective but short, sharp messages which do not require specific training or significant time.

The potential benefits of this increased reef visitation for conservation are, however, extremely important. Tourism is economically significant and its revenue can contribute to conservation through direct spending, such as visitor fees to parks or donations. Park fees, if set and collected effectively, have the potential to provide significant income to MPA managers. Although the marine environment has the difficulty of being unfenced and having open access, most tourism activity on the ocean takes place via the 'fence' of private operators. This means that fee collection cannot usually take place at a traditional terrestrial park-style gate but that private operators can act as park officers in collecting fees. A close cooperative relationship is required between park authorities and the private sector but such a fee mechanism can help to create and maintain that relationship. The same public/private relationship issue applies to the use of environmental education, as discussed later.

Often more significant economically is tourism's potential for providing revenue to local communities. Employment, an increased market for the sale of goods and new business and training opportunities can result to local people from tourism. Where the declaration and enforcement of a protected area takes place, alternative livelihood options become fundamental to the success of conservation objectives. That local communities should benefit is a fundamental tenet of sustainable tourism but the opposite scenario has been too often the reality. Ownership and profit making by expatriates or local elites has been a common model of tourism in the developing world. The introduction of tourism to an area requires significant investment in infrastructure, marketing and staff. This very often comes from outside the area and ownership and power remain in the hands of outsiders. In marine tourism, this is especially significant, since many activities require greater training and equipment than in terrestrial natural areas and expatriate ownership of companies is often the norm. In the dive industry, specifically, the costs associated with instructor training and equipment mean that expatriate ownership and employees are extremely common in developing countries. Marine tourism, simply by virtue of the equipment, skills and boats or other craft necessary for most activities, implies that the involvement of poorer people often requires genuine effort and pro-active initiatives on the part of governments, protected area authorities and the tourism industry. We have begun to see this happening in community-based tourism initiatives on land, but have yet to see significant progress from the major players in the marine tourism industry. There has been significant work and investment on the part of agencies like PADI to address environmental conservation issues but the social responsibilities of such a large industry as scuba diving have yet to be accepted and addressed. Community benefit, however, is partly the responsibility of MPA managers, not only morally but also to maintain the good relationship necessary with one's neighbours for success in conservation.

Tourism's role as revenue provider, then, is potentially important to managers but the significant numbers of tourists now experiencing the marine environment also have a wider potential for awareness raising and public interest in and support for reef conservation. Public relations, broadly speaking, is a consideration for many MPA managers. If the MPA model is to be successful and genuine, and not just 'paper parks' created in marine and coastal areas, international and national organisations and governments with the legislative and economic power to create and fund MPAs need to see successful examples.

Equally, or perhaps more importantly, such areas need local support. Work on the fishery benefits of MPAs is ongoing (Raloff, 2001) and is producing evidence that increased fish stocks in surrounding waters result from the existence of MPAs, where fishing is either not allowed or is restricted. However, obtaining this evidence, and indeed showing the visible fishery benefits of MPAs, is a relatively long-term process; restrictions on fishing and other extractive uses are often required in the meantime. Tourism is an important option, where viable, for alternative economic activity that can produce results relatively quickly. Although it is over-simplistic to see tourism as a straightforward and complete replacement for an industry such as fishing, it does have a contribution to make in a number of ways. In a number of terrestrial parks, historical experience has been that of exclusion of local people from their traditional natural resources and livelihoods, turning hunters into poachers who are expected to abide by park rules while watching outsiders – tourists – use those resources for recreation. It is essential that MPAs do not repeat this experience and that tourism is not and is not seen to be allowing privileged outsiders to replace one extractive activity with another. Tourism on reefs must be genuinely non-extractive and environmentally sustainable, in order to be socially sustainable: the support of people and organisations who have a stake in it is necessary for its economic sustainability.

Education, Interpretation, Propaganda . . . Changing Tourists' Behaviour

Education is a useful management tool in addressing tourist impact in MPAs. It can, if done well, provide the necessary bridge between tourism and conservation that creates the symbiotic relationship between the two. It can also create support for marine conservation both among the tourists who experience it and among funders, legislators and local communities.

Environmental education can also be free or at least relatively cheap. Interpretation in terrestrial parks can be expensive and may require significant maintenance. Most of the activities that make up marine tourism make this type of interpretation more difficult, since the time available to address environmental issues with marine tourists is often restricted to that on board a boat. This restricts time and space,and issues such as weather and seasickness make sophisticated interpretation difficult.

This situation does have its advantages, though. A boat provides a genuinely captive audience: since they generally cannot escape, tourists are likely to attend to messages. Since the marine environment is, for most, unfamiliar, tourists are also more likely to listen to instructions and recommendations attentively, for safety reasons.

As mentioned previously, most access to the ocean in marine tourism involves the private sector and this can be turned to the advantage of the MPA manager. If private operators taking divers, snorkellers, whalewatchers and so on can provide good, effective environmental information, they can do a job for the usually over-stretched manager and add value to the experience they provide. Good relationships between protected area authorities and private operators are important for managerial success and can be promoted by collaboration in education.

Commentators and practitioners use a number of words and disciplines to describe the activities that attempt to inform and inspire tourists. 'Education' tends to imply a relatively formal activity, often taking place in the classroom with a group of people expecting to learn. 'Interpretation' is the discipline more commonly employed to encourage appreciation and understanding of natural or cultural sites, as well as to control visitor impact. Environmental psychologists also work on ways of changing human behaviour to protect or improve the environment. Ultimately, these mechanisms can be applied to tourism and be used for the same objective: attitude and behaviour change among tourists.

Despite the separate nature of these disciplines, they often arrive at similar conclusions. However, the experience and knowledge gained by professionals is not always shared, and operators or conservation workers on the front line of marine tourism rarely have the information and training that would be helpful in making their efforts successful. Evaluation of educational activities is also notably lacking, so that ineffective practices may continue for years with good intentions but insignificant results.

While success in achieving change can be fairly simple, it is not necessarily something that can be achieved instinctively, even when there is extensive factual knowledge, dedication and enthusiasm on the part of those delivering it. Information provision based on tried and tested methods, shared among practitioners, is far more effective than individuals working in separate bubbles to achieve the same goals.

Although environmental psychology and interpretation are, in many ways, separate, they can both be used for the same purpose, that is, securing behaviour change. It is perhaps surprising just how separate the two disciplines of interpretation and environmental education have become.

For those concerned with interpretation, Tilden's book, *Interpreting Our Heritage,* is generally considered to be the original 'bible' of the field. It defines the philosophy of the art or skill of interpretation as 'the work of revealing, to such visitors as desire the service, something of the beauty and wonder, the inspiration and spiritual meaning that lie behind what the visitor can with his senses perceive.' (Tilden, 1977: 4) Others have followed Tilden with practical guides to interpretation, with advice and examples of its various tools, guided tours, signs, brochures, audio-visual exhibits and so forth (Ham, 1992; Sharpe, 1982). The profession itself has developed and professional societies and journals exist in many countries.

Conversely, environmental psychology, or that branch of it which is concerned with changing people's environmental behaviour, takes a much more theoretical and experimental approach to the problem. Theorists and practitioners rarely cross paths in the field of visitor management to natural areas.

Environmental psychology has looked at methods for changing attitudes and behaviour. Research has shown that the link between environmental attitudes and behaviour is complicated. Despite early evidence that pro-environmental attitudes had little to do with pro-environmental behaviour, the consensus now, based on recent research, is that the two are related (Bell *et al.*, 1996; Gifford, 1987; Veitch & Arkheim, 1995). This link, however, is not specific. More recent research has shown there to be a correlation between a person's overall environmental attitude and the number of pro-environmental activities they take part in, but not necessarily that an attitude translates directly into a behaviour that might be considered compatible with that attitude. As Bell *et al.* (1996: 335) put it, 'a pro-environmentalist may not keep the thermostat at 65 degrees in the winter, but someone who adheres to several pro-environmental concepts probably does engage in more pro-environmental behaviours . . . than someone who is not concerned with the environment'.

On the subject of how to change environmentally relevant attitudes and behaviour, psychology and interpretation have separately reached a similar conclusion. Although psychologists do have a use for formal education, neither they nor interpreters consider this the best way to communicate an environmental message. For psychologists, pure information has its place with children and with those who may have compatible pre-existing attitudes but 'are simply lacking in relevant knowledge' (Veitch & Arkheim, 1995: 428). So simply providing factual information is only likely to be effective when 'preaching to the converted'. If the audience is likely to agree with the message they are hearing they only need to be given specific instructions to change a particular behaviour. For example, a group of environmentally-committed people who generally live a 'green' lifestyle, recycle waste, minimise their energy use, buy organic products and so on, may know nothing about how to protect coral reefs. This audience may need only to be given relatively dry information about the ecological importance of reefs, in order to modify their behaviour.

In general, however, research tends to agree that simply giving people information is considered the least effective way of affecting their attitudes and behaviour. Psychology provides scientific evidence for Tilden's assertion that 'the chief aim of interpretation is not instruction but provocation' (Tilden, 1977: 32).

Various alternatives to information have been tested by psychologists and are briefly outlined here:

- Behavioural psychological experiments that use reward and punishment mechanisms for 'good and bad' behaviour have been carried out, especially with litter. Although these can be quite effective, dependence on rewards is rarely practical in the long term (Gudgion & Thomas, 1991).
- Psychologists recommend appeals to the 'affective domain', which relates to a person's emotional feelings towards a place or other entity. This also tallies with Tilden's philosophy.
- Festinger's theory of cognitive dissonance and Piaget's work on disequilibrium that followed are much quoted in this field (Forestell, 1990; Gudgion & Thomas, 1991; Orams, 1995). The theory of cognitive dissonance states that new information will lead to either 'assimilation', if it agrees with a person's pre-existing beliefs and knowledge, or 'accommodation' if it challenges those beliefs and must therefore be 'accommodated' by a change in attitude/belief and behaviour. It is a combination of assimilation and accommodation that leads to learning and behavioural change, although accommodation will not take place if it is too drastic, that is, too far outside the realms of pre-existing beliefs.

Combining Theory with Visitor Management

> [D]espite a significant amount of scientific effort directed at understanding how people learn, little of this information has been directed at models which can make environmental interpretation more effective. (Orams, 1994: 32)

The obvious obstacle to environmental education for tourists is that their primary motivation is often entertainment and relaxation. It is sometimes assumed that any behavioural guidelines or the serious, restrictive messages that tend to be associated with environmental conservation will be contrary to their aims and interests. Although this is a simplistic assumption, assuming the stereotypical contradiction between purist, do-gooder greens and hedonistic sun, sea and sand seekers, formal education is not an option in all but a small niche of the tourism industry. Tourists are described by interpreters as a 'non-captive' audience (that is voluntary and not necessarily attentive) and they often display very different levels of knowledge and interest, as well as varying greatly in factors such as age and education.

A message needs, then, to address the whole, diverse audience of visitors or participants and to achieve the desired management objective of attitude and behaviour change without being didactic, patronising, lengthy or boring. That is not an easy job.

It may be, however, that as Orams (1994) points out, wildlife tourism has an advantage for the effectiveness of interpretation. It can appeal to the emotions or 'affective domain' and presents an opportunity, through the new and perhaps strange behaviour of animals, to create cognitive dissonance. The opportunity to appeal to these two psychological processes may create 'short-cuts which are able to be used to counter the problems inherent in educating tourists' (Orams, 1994: 32). Orams describes an experiment where an interpretation programme for a dolphin-feeding experience was designed according to the theories outlined earlier. This was compared with the existing programme by testing tourists' attitudes and behaviour. The experiment found a significant, long-term improvement in respondents' environmental behaviour (joining a conservation organisation, picking up litter that might harm marine mammals) in those people exposed to the new programme based on psychological learning theory.

Those working in marine tourism also have on their side the advantage of the positive effects of direct experience. Experiencing something first-hand has been shown to be more effective in changing behaviour than 'abstract' learning. The normally shaky link between attitude and behaviour becomes much stronger when combined with direct experience (Manfredo & Bright, 1991; Veitch & Arkheim, 1995).

Passive experience though is not enough: tourists need to become involved directly through their own actions, and there needs to be what psychologists call 'opportunity to act'. As Brylske (2000: 4) puts it, '[j]ust because the interpretive experience is personal does not necessarily mean it is effective'. Opportunity to act is the point at which attitude and good intentions become behaviour. It is essential that tourists are given some way of acting immediately on the information they have been given to cement their knowledge and opinion into action.

Assumptions are often made instinctively that are not backed up by research findings. The most obvious, discussed earlier, is that 'awareness raising' or simply disseminating information, will lead to attitude and behaviour change. Another is that 'it is very easy to interpret to the educated and motivated elite. It is, however, vital that the difficult task of interpretation for the mass audience with little time or interest be tackled' (Bramwell & Lane, 1993: 78).

This is backed up neither by psychological theory nor the evidence of diver impact research. Psychologists' work showing that it is far easier to affect beliefs and behaviour if a person holds no strong previous views (Veitch & Arkheim, 1995) can explain findings that experienced divers had no smaller effect on the environment than novices, despite their superior skills, and that photographers (who are generally experienced) and relatively expert divers are the most damaging single groups. Manfredo and Bright's (1991) work has shown that the higher the level of a person's previous knowledge, the greater their resistance to attitude change and new information. They also show that it is far easier to affect a new or inexperienced visitor by means of interpretation than an expert.

This is not hard to believe when one thinks of, for example, experienced scuba divers who considers themselves to be knowledgeable and relatively expert, compared with new divers who are unsure of the environment and their own skill in it. 'Experts' are less likely to listen to guidelines than are slightly nervous novices. Expert types also have a tendency to consider rules and regulations to be irrelevant to them. The logic that dictates that one person will not make a difference to the overall state of the reef applies here. Divers, birdwatchers or keen gardeners who consider themselves to be experts may also consider themselves to be above environmental guidelines, which they see only as relevant to 'the masses who cause the real damage'.

This problem of the 'expert' visitor has important ramifications for interpreting natural sites for tourists, especially divers and snorkellers. It means, despite the difficulties, that it is important to target the knowledgeable visitor at least as carefully as the new. In snorkelling, and especially diving, the training and skills required are likely to attract higher numbers of repeat visitors, if not to the same site then to coral reefs in general. These regular divers, having invested time and money in their hobby and developed an interest in the activity, are perhaps more likely to become more knowledgeable than the average visitor to a terrestrial park. These people are likely to be psychologically less well disposed to listening to and assimilating information, but to have the same if not greater negative impact on the reefs as the easy-to-influence novices. It is therefore important to address education and interpretation to them. In diving, among other activities with a natural focus, there is a disproportionate number of expert types, and this is likely only to grow with the number of divers becoming certified.

Traditionally, the difficulty of addressing such diverse audiences has been addressed by interpreters by attempting to 'layer' their message. Signs for example, have different levels of information on them, ranging from basic to detailed, so that all audiences are catered for. This relies on the presumption that the more knowledgeable expert visitor will want and will take the time to seek out relevant new information rather than, in fact, being less open to messages and information than the new visitor. It also assumes once again that information provision is paramount.

Diver impact studies have shown that, in various countries, there is a conflict between recreational use (and its consequent economic benefits) and the conservation objectives of MPA managers. It is known than divers and snorkellers have a significant negative environmental impact on the coral reefs that attract so many tourists. Education can reduce that impact if designed carefully.

Environmental education, psychology and interpretation have shown that there is more than one way to educate tourists and thereby to affect behaviour and reduce impact. There are some very significant points on which all disciplines broadly agree and which should be considered in the design of educational programmes, materials or briefings:

- Knowledge does not directly affect attitude and attitude does not necessarily affect behaviour.
- Messages should be short and not overload on information. Five pieces of information are considered the average maximum that an audience can take in.
- People should be given the opportunity to carry out good intentions. It is not enough to tell people to do or not to do something: they need to be given the tools or skills to do it and an opportunity, ideally immediate, to carry it out.
- Cognitive dissonance or causing people to re-think their opinions can be effective in attitude and behaviour change. This may be helped by creating questions in people's minds, which are then answered.
- Appealing to people's emotions, by discussing human and emotional topics like birth and death can be more effective than dealing in dry or abstract facts.

A Case Study of the British Virgin Islands

A study designed to test the validity of these ideas was carried out in the British Virgin Islands and is described here.

The dive instructors who participated in this study had a generally high level of environmental knowledge and commitment, and often included information on reefs and their creatures in their dive briefings. These were dependent on the knowledge of whoever was doing the briefing, the time available and the perceived knowledge and interest of the day's divers. It tended to be pure information giving: the most common form of information was the naming of fish species likely to be seen on that site (see Figure 8.1). Dive groups on any given boat tend to have widely varying levels of knowledge of fish, which means that species-naming will inevitably bore some who consider the level too basic and confuse others who have never heard of any of them. This type of briefing was also unlikely to affect divers' commitment to conserving the reef through their own actions.

Information on divers' knowledge of reef species and environmental threats was collected through questionnaires and was compared with the diver's impact on the reef. Impact was tested in the standard way, that is to observe divers (without their knowledge) for a 10-minute period and record the number of times they contacted the reef, deliberately or accidentally.

Despite dealing with poor skills and ignorance every day, and being aware of the damage caused by these factors, no guides involved in this survey gave behavioural instructions regarding the environment as a matter of course. Divers were sometimes asked not to wear gloves and to replace shells brought to the surface but briefings did not include simple guidelines relating behaviour to environmental damage. Perhaps it was assumed that all divers know that touching the coral damages it and so staying away from it avoids that. The majority probably do know this but it seems they need to be reminded. Guides were therefore asked simply to

Figure 8.1 Identification photo for typical reef fish

include three short, sharp messages in their normal briefing. These points echoed information on posters that were put up on board the dive boats:

- that coral is an animal and is fragile (see Figure 8.2);
- the 'magic metre' rule that asks divers to make sure they remain at least 1m off the bottom at all times to avoid kicking coral or kicking up sand; and
- horizontal body positioning, to avoid unconsciously kicking the bottom. If a diver is positioned vertically or diagonally, it may feel as though they are a good distance above the reef. However, the diver's fins are likely to be hitting the bottom or the branching coral that grows up from it (see Figure 8.3).

Each guide was asked to include these points as short messages, even if other environmental information was given. They deal with the most common causes of diver damage: ignorance, poor buoyancy and poor body positioning.

Instead of or as well as pure information, this change attempted to reach divers differently. The questionnaires had shown that knowledge of reefs was surprisingly basic and that by no means the majority of divers would know that coral is an animal rather than a plant. This fact, and that it is fragile and vulnerable to our actions, appeals more to divers' emotional side than naming species. The other two points both fall into the 'opportunity to act' category: they provide divers with an immediate and easy tool to ensure they do nt hit the reef. Again, these are fairly

Answer: It is all three at once.

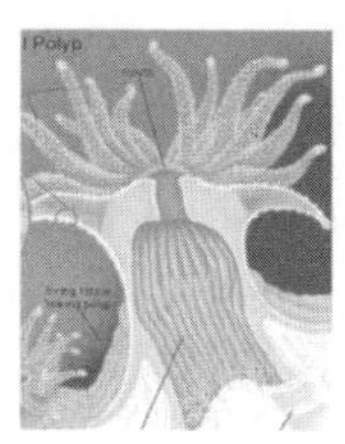

The picture shows a magnified coral polyp, one of the tiny, fragile animals which make up coral colonies.

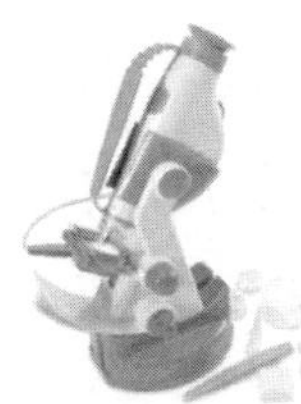

Microscopic algae called zooxanthellae live symbiotically inside the polyps, these plants photosythesise and feed their animal hosts. They also give coral its colour.

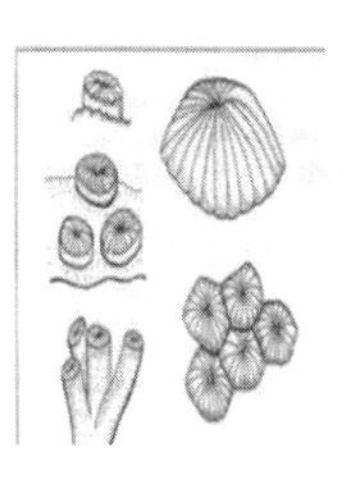

Together, these two make calcium carbonate, the mineral which forms their protective skeleton and the reef structure that we see.

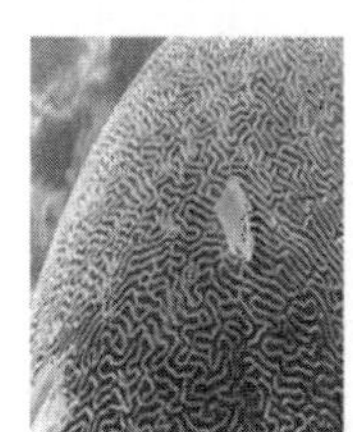

So, each coral colony that you see is a complex combination of animal, plant and mineral. They are fragile and easily damaged by our touch so please be aware and avoid touching live coral.

You are a customer on the dive boat but a guest on the reef.
Please treat it with respect and leave it as you would wish to find it.

Figure 8.2 Briefing poster on coral species

How to be an Eco-Diver

Consider these recommendations for no-impact diving and make your dives easier, more fun and reef-friendly. Play your part in protecting coral reefs and help to ensure that they are here for you and your children's future pleasure.

Buoyancy
Follow the 'Magic Metre' rule. Stay 1m (3ft) above the bottom to avoid kicking corals or sand, which can suffocate coral polyps.

Positioning
Stay horizontal whenever possible. If you need to rest on the bottom, choose sand.

Equipment
Keep gauges and other dangly bits secured close to the body, they'll be easier to find and won't hit living creatures.

Marine Life
Observe from a happy distance, the creatures will stick around if you don't get too close. Chasing, feeding or removing anything can upset the reef's balance and harm the wildlife.

Learn
The more you know about what you see, the more enjoyable diving becomes. Ask staff about what you see, use ID slates and books, think about a specialty course such as reef ecology.

Get Involved
There are are many things you can do to help, be a reef ambassador to other divers, snorkellers, friends and family. You could be one of the many people worldwide, from beginners to experts, who take part in surveys of reef health. Check out the organisations shown here or ask the dive staff.

IF YOU ARE UNSURE
Ask! Your guide can help with buoyancy tips or even a specialty course. If you need to hold on, in a current for example, they can tell you where you can touch without causing harm. If you want to know about a creature they can identify it or help you do so. If you want to get involved, they can point you in the right direction.

Be proud to be a diver. Enjoy the reefs & avoid damaging the life you dive to see.

Figure 8.3 Briefing poster on diving good practice

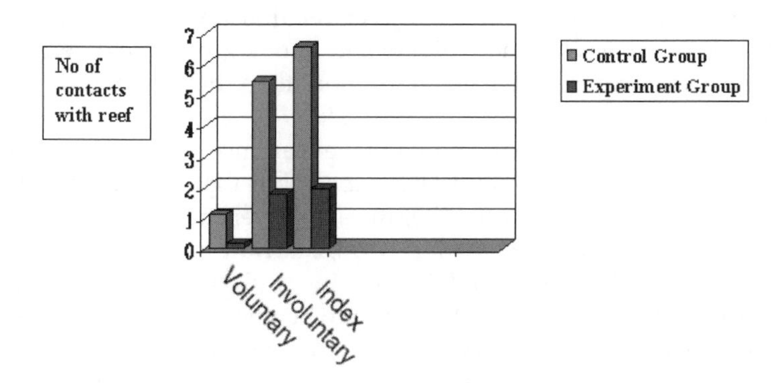

Figure 8.4 Contacts with reef for control and experiment groups, showing the effect of the modified dive briefing

obvious actions but may not have been pointed out to divers in the past or they may need reminding of them regularly.

Whether as new ideas or reminders, these briefing points succeeded in changing divers' behaviour and so in reducing the damage they were causing.

Figure 8.4 shows the drop in the number of times divers touched, banged into or otherwise contacted the reef. It divides those contacts into voluntary and involuntary, and shows the control group, divers who heard the usual variety of briefings, compared with the experiment group, those who heard the three points outlined earlier. Very few people deliberately touched coral or other animals even in the control group – it was accidental contacts, fins kicking corals or divers getting too close to the bottom, that were causing significant damage. It seemed that most divers knew they should not be grabbing coral or harassing animals but that they needed to be reminded of the importance of these and to be given easy immediate tools to let them carry out their intentions.

Some interesting results came from the correlation of these touch rates and other variables. Divers' experience, environmental knowledge and commitment as well as factors like gender and age were compared with their impact.

Both groups knew relatively little about local fish and reefs, the average score being between three and four out of five. The divers did know more about global threats to reefs, scoring an average of almost four out of five on those questions, which is encouraging.

Divers who knew a lot about local reefs might have been assumed to be those who had dived there more often and who had learnt about fish and their habitats because of an interest in ecology. This was not borne out by the results, however. Divers' knowledge level made no difference to their impact, so a diver who knew a lot about local fish did not necessarily touch the reef less often.

There was a correlation between diving experience and reef contacts. Overall, the more experienced divers tended to touch the reef slightly less often than relatively inexperienced divers. This was not a direct relationship, however. Complete beginners had less impact than new divers and in one group the divers with the most experience, the 'experts', touched a lot more than those with fewer dives in their logbooks.

Some previous studies have also shown that experience, and the improved skill that result from this, do tend to minimise impact; but that novices tend to show greater caution (indeed, a substantial literature on diver impact exists: for greater detail see Allison, 1996; Davis & Tisdell, 1995; Harriott *et al.*, 1997; Hawkins & Roberts, 1992a, 1992b, 1993, 1994, 1997; Hawkins *et al.*, 1999; Rogers *et al.*, 1988; Rouphael & Inglis, 1995; Talge, 1992). Complete beginners also tend to be diving with an instructor who is likely to be exercising greater than normal caution and control over that diver. This implies that closer monitoring of inexperienced divers as well as greater concentration on skill development during diver training could have an additional positive effect on diver damage.

The only other variable that showed any relationship with environmental impact was gender. In both groups women touched the reef significantly less often than the men studied. Management options based on that finding might be a little more controversial than modifying dive briefings.

Conclusions

This chapter has looked specifically at dive tourism and one experiment with modifying dive briefings. Although relatively modest in scope, it provides encouraging evidence that very simple tools can make a significant difference to reef damage by tourists. A few other studies in marine tourism have been equally encouraging and work continues to test a greater variety of educational methods on divers in Indonesia (Jensen, pers. comm.). Since significant research, practice and evaluated evidence exists which can be used to design or modify education for marine tourists, it is important to use that and progress, rather than reinvent the wheel. Enough evidence exists that we need only to modify this wheel for the water, rather than try to re-design it.

In dive tourism, there is a notable lack of environmentally relevant training throughout a diver's career. It is essential that instructors and dive masters be given the tools most are keen to use that allow them to get their often genuine passion for the marine environment and its preservation over to students and customers. Currently, environmentally specific training tends only to happen when a diver chooses to spend time and money on a formal course. It needs to be integrated into all training and into day-to-day operations of dive operators and marine park staff.

Although other marine activities will have different contexts, the same principles apply. It is to be hoped that by using them and by evaluating our success wherever possible, change can be achieved.

Where the diver tourism industry succeeds in using education to reduce or remove environmental damage caused by tourism and to increase public support for marine conservation, two very important steps towards genuinely sustainable tourism have been taken. Marine parks and those with a stake in them have an opportunity to use and learn from the mistakes and successes of terrestrial ecotourism. It would be a great waste of extremely limited time and resources not to use that experience. Education can be implemented quickly and cheaply and there is plenty of adaptable experience available.

References

Allison, W.R. (1996) Snorkeller damage to coral reefs in the Maldive Islands. *Coral Reefs* 15, 215–16.

Bell, P.A., Greene, T.C., Fisher, J.D. and Baum, A. (1996) *Environmental Psychology*, 4th edn. Fort Worth: Harcourt Brace Publishers.

Bramwell, B. and Lane, B. (1993) Interpretation and sustainable tourism: The potential and the pitfalls. *Journal of Sustainable Tourism* 1 (2), 71–80.

Brylske, A.F. (2000) *A Model for Training Tourism Professionals in Tropical Marine Resource Management*. Cape Coral, FL: Instructional Technologies.

BSAC (2001) Information sheet A.15. Unpublished document, The British Sub-Acqua Club, Ellesmere, Cheshire.

Davis, D. and Tisdell, C. (1995) Recreational scuba-diving and carrying capacity in MPAs. *Ocean and Coastal Management* 26 (1), 19–40.

Forestell, P.H. (1990) Marine education and ocean tourism: Replacing parasitism with symbiosis. In M.L. Miller and J. Auyong (eds) *Proceedings of the 1990 Congress in Coastal and Marine Tourism*, Vol. 1. Corvallis: National Coastal Resources Research Institute.

Gifford, R. (1987) *Environmental Behaviour: Principles and Practice*. Neeham Heights, MA: Allyn and Bacon.

Gudgion, T.J. and Thomas, M.P. (1991) Changing environmentally related behaviour. *Environmental Education and Information* 10 (2), 101–12.

Ham, S. (1992) *Environmental Interpretation: A Practical Guide for People with Big Ideas and Small Budgets*. Boulder, CO: North American Press.

Harriott, V., Davis, D. and Banks, S.A. (1997) Recreational diving and its impact in Marine Protected Areas in Eastern Australia. *Ambio* 26 (3), 173–9.

Hawkins, J. and Roberts, C.M. (1992a) Can Egypt's coral reefs support ambitious plans for diving tourism? *Proceedings of the Seventh International Coral Reef Symposium* (pp. 1007–13). Mangilao, Guam: University of Guam Press.

Hawkins, J. and Roberts, C.M. (1992b) Effects Of recreational scuba-diving on fore-reef slope communities of coral reefs. *Biological Conservation* 62 (3), 171–8.

Hawkins, J. and Roberts, C.M. (1993) Effects of recreational scuba-diving on coral reefs: Trampling on reef-flat communities. *Journal of Applied Ecology* 30 (1), 25–30.

Hawkins, J. and Roberts, C.M. (1994) The growth of coastal tourism in the Red Sea: Present and possible future effects on coral reefs. *Ambio* 23 (8), 503–8.

Hawkins, J. and Roberts, C.M. (1997) Estimating the carrying capacity of coral reefs for scuba-diving. *Proceedings of the Eighth International Coral Reef Symposium* (pp. 1923–26). Balboa, Panama: Smithsonian Tropical Research Institute.

Hawkins, J., Roberts, C.M., Van't Hof, De Meyer, K., Tratalos, J. and Aldam. C. (1999) Effects of recreational scuba-diving on Caribbean coral and fish communities. *Conservation Biology* 13 (4), 888–97.

Manfredo, M.J. and Bright, A.D. (1991) A model for assessing the effects of communication on recreationists. *Journal of Leisure Research* 23 (1), 1–20.

Orams, M.B. (1994) Creating effective interpretation for managing interaction between tourists and wildlife. *Australian Journal of Environmental Education* 10 (2), 21–34.

Orams, M.B. (1995) Using interpretation to manage nature-based tourism. *Journal of Sustainable Tourism* 4 (2), 81–94.

PADI (2001) Online documention [http://www.padi.com/news/stats/].
Raloff, J. (2001) Underwater refuge. *Science News* 159 (17), 264–6.
Rogers, C.S., McLain L. and Zullo E.S. (1988) Damage to coral reefs in Virgin Islands National Park and Biosphere Reserve from recreational activities. *Proceedings of the Sixth International Coral Reef Symposium* (pp. 405–10). Townsville.
Rouphael, T. and Inglis, G. (1995) The effects of qualified recreational SCUBA divers on coral reefs. CRC Technical Report. Townsville: James Cook University of North Queensland.
Sharpe, G.W. (ed.) (1982) *Interpreting the Environment*. New York: John Wiley and Sons.
Talge, H. (1992) Impact of recreational divers on scleratinian corals of the Florida Keys. *Proceedings of the Seventh International Coral Reef Symposium, Guam* 5 (2), 1077–82.
Tilden, J. (1977) *Interpreting Our Heritage*. Chapel Hill: University of North Carolina Press.
Veitch, R. and Arkheim, P. (1995) *Environmental Psychology*. New Jersey: Prentice Hall.

Chapter 9
Reconciling Communities' Expectations of Ecotourism: Initiating a Planning and Education Strategy for the Avoca Beach Rock Platform

MATTHEW McDONALD and STEPHEN WEARING

Introduction

Australia's coastal and marine environment has come under enormous pressure in the last 40–50 years from urban development, pollution, invasive weeds and marine pests, as well as human use and abuse in the form of tourism and recreation. Many coastal communities in Australia have watched their precious marine environments degrade quickly with each subsequent holiday period. The significance of this issue is all the more alarming for coastal communities situated close to Australia's major metropolitan centres. One particular marine environment undergoing change as a result of tourism is the Avoca Beach Rock Platform. To stem the increase in physical and social impacts, a series of meetings was held between Coastcare (a federal government agency), researchers and the Avoca Beach community in order to develop some strategies to conserve the local environment.

These meetings highlighted the need for visitor education as part of an overall management strategy. The issues facing Avoca Beach could clearly benefit from a framework based on the tenets of ecotourism. To facilitate this strategy, a study was carried out to ascertain the visitor patterns and characteristics associated with Avoca Beach Rock Platform. Visitors were asked a range of questions relating to demographics, the type of visit and strategies to manage the rock platform better.

The results indicate that education needs to be targeted at line fisherman. This can be achieved through the development of a community-directed education programme. This chapter argues that this specific strategy needs to be complemented by a region-wide tourism and recreation plan, underpinned by the aims and philosophies of ecotourism.

Ecotourism and Communities

A significant contribution to ecotourism's global following has been its potential to deliver benefits to communities. However, it is essential that an assessment is made of how, at site level, local communities regard this contribution. In the past, it has been the dominant economic focus that has served to obscure significant dimensions of tourism impact. Tourism produces a diverse range of both social and environmental impacts that are often complex and interrelated. One local representative put it in simple terms: 'Tourism is like fire. It can cook your food or burn down your house' (cited in Fox 1977: 44). The tourism industry makes extensive use of natural assets – forests, reefs, beaches and parks – but what does it contribute to management of these assets? Both the provision of tourism infrastructure and the costs of managing the impact of tourism on host communities are often borne by the environment, the community itself and the government. A significant body of research has challenged the claims of industry and government agencies that the aggregate benefits of tourism far outweigh the costs: 'benefits are rarely uniform, accruing to those actively involved in the tourist industry, while costs are often borne by those who derive no compensatory benefits from tourism' (Butler, 1980: 572).

Local communities are significantly vulnerable to the deleterious impacts of tourism development as they directly experience the socio-cultural impacts of tourism. The subsequent impact of tourism's dynamic growth on communities has, in some cases, precipitated strong protests by community groups which, being sensitive to the impacts of tourism, have actively opposed large-scale tourism developments for their locality. Other community groups have been more accepting of a gradual growth in tourism to their region over many years, only to become aware of the negative impacts at a later date when these impacts cannot easily be ignored.

The tourism potential of local areas is also compromised by the environmental impact of other industries. According to the Economist Intelligence Unit, the entire tourism industry is under attack from other business interests that are virtually stealing its assets (Jenner & Smith, 1991). In the late 1980s, the development boom initiated the emergence of many so-called 'tourism developments' which were nothing more than land speculation or a means of making otherwise conventional residential developments acceptable to planning authorities. It led to bankruptcies, inflated profits, overloaded infrastructure, residential sprawl and unwanted social and environmental impacts, which led to many local communities becoming highly suspicious about the benefits of the tourism industry. The ecological, cultural and social impacts of tourism often lead to diminished community and political support for the industry, particularly at local levels.

The interdependence of tourism and the social and physical environment is fundamental to the future of each, and seeking a way to accommodate the needs of all parties without control being external to those who experience its effects most directly is essential. Features of the natural and cultural environment and supportive host communities are the foundations of a successful industry. Neglect of conservation and quality of life issues threatens the very basis of local populations and a viable and sustainable tourism industry.

Ecotourism has the potential to create support for conservation objectives in both the host community and in the visitor alike, through establishing and sustaining links between the tourism industry, local communities and protected areas. As social and environmental benefits are essentially interdependent, social benefits accruing to host communities as a result of ecotourism may have the result of increasing overall standards of living, due to the localised economic stimulus provided for in increased visitation to the site. Similarly, environmental benefits accrue as host communities are persuaded to protect natural environments in order to sustain economically viable tourism (Ceballos-Lascuraín, 1990).

Many tourists, especially ecotourists, are sensitive to decreases in water quality and air quality, loss of vegetation, loss of wildlife, soil erosion and a change in the character and visual appeal of an area due to development. Degradation of the natural environment will severely reduce visitor demand in the long term, because the natural attributes upon which ecotourism demand is based will be perceived as less attractive, less legitimate and less able to provide satisfying ecologically-based experiences.

There are a number of reasons why ecotourism may be considered by local communities:

- a desire to be part of strong growth in tourism generally and recognition of the potential of catering for special interest tourism (niche markets);
- an awareness of the high value of natural attractions in the locale;
- empathy for conservation ideals and the need for sustainable tourism; and
- a desire to rejuvenate the local tourist industry in a responsible manner.

As these benefits suggest, ecotourism is about attracting visitors for the 'right' reasons and not simply the promotion of tourism for the sake of the 'tourist dollar' at the expense of a community's natural and cultural attributes. However, local communities are not immune from ecotourism impacts.

The conflicting issues expressed by representatives of host communities to tourism development generally fall into a number of interrelated categories:

- the lack of opportunities for involvement in decision-making relating to ecotourism;
- inadequate responses from governments when administrative or legislative mechanisms have been established to involve them in such decision-making;

- the lack of financial, social and vocational benefits flowing to these communities from projects that exploit commercially that which they regard as their resources;
- the need to establish better tools for evaluating socio-cultural impacts and for ensuring this is completed in addition to the more often emphasised impacts on the natural environments, which are usually of more interest to the outside investors and conservation groups;
- impacts on community cohesion and structure; and
- the rapidity of tourism development that, in many cases, accelerates social change significantly.

These concerns embrace a wide range of issues relating to the management of natural resources adjacent to these communities. The central issue is the inadequate levels of participation perceived by these communities in the management of what they regard as their domains.

The Avoca Beach Rock Platform

Avoca Beach and the Avoca Beach Rock Platform are located 16 km east of Gosford on the Central Coast of New South Wales, Australia. The area is a two-hour drive north of metropolitan Sydney via the Pacific Freeway. The beach and rock platform is a valuable coastal retreat for fishing, swimming, snorkelling, collecting, exploring and relaxing. As well as providing for these recreation activities, the area is home to organisms such as algae, sea anemones, abalone, limpets, barnacles, elephant snails and a wide variety of marine life.

Like many coastal environments close to large population centres, the Avoca Beach Rock Platform suffers high impacts from large volumes of tourist visitation. During peak holiday periods, such as Christmas and Easter, large amounts of rubbish are left on the platform, particularly rubbish related to fishing activities such as fishing line, hooks, floaters, sinkers, bait bags and raw bait. Illegal activities such as camping on the platform, the collecting of intertidal organisms, the taking of undersized fish and vandalism are all issues that are creating conflicts within and between different user groups and the local community adjacent to the rock platform.

Tourism and environmental impacts have frustrated and angered the local community, particularly the Avoca Beach Progress Association and the Avoca Beach Surf Life Saving Club. Many in the local community feel the problem is caused by ethnic fishermen from the outer suburbs of Sydney. It is alleged that these groups travel from Sydney to fish from the rocks, camp overnight on the rock platform, over-collect the platform flora and fauna and discard their rubbish when they leave.

This distressing situation is not unusual in Australia or overseas (Heiman, 1986). Coastal visitors can be viewed as either a community mainstay or a bane:

> Because coastal recreation and tourism are so popular, residents often see themselves as victims of visitor abuses ... Although local residents often emphasise the adverse effects of coastal recreation and tourism, coastal communities can also reap substantial economic and social benefits from the tourist trade. (Heiman, 1986: 34–5)

The Australian National Ecotourism Strategy

In 1994, the Federal Government of Australia established a National Ecotourism Strategy with the intention of formulating an overall policy framework for the planning, development and management of ecotourism to contribute to the achievement of sustainable tourism in natural areas (Evans-Smith, 1994). Such a strategy highlights the importance of the government's role in establishing the necessary guidelines through which ecotourism can be developed in accordance with the principles of sustainability. Through the formulation of a broad framework promoting sustainable ecotourism, the federal government potentially had a great amount of influence in determining the future direction of this type of tourism and thus its sustainability. However, since the change of government in Australia following the federal election of 1996, the National Ecotourism Strategy has not been promoted or recognised as a valid strategy by the newly elected government. In this instance, the government support in implementing this strategy has been withdrawn and leaves the impetus of any further actions with the industry and state governments. Some states in Australia have developed their own ecotourism or nature-based tourism strategies.

To ensure that ecotourism is ecologically sustainable, the Commonwealth Department of Tourism emphasises the terms 'education' and 'interpretation' of the natural environment in the National Ecotourism Strategy. This is because education and interpretation are seen as significant tools in increasing enjoyable and meaningful ecotourism experiences and assisting in the sustainability of tourist activities in natural areas (Queensland Department of Tourism, Sport and Youth, 1995).

This recognition of the centrality of interpretation and education as core components of ecotourism helps to differentiate ecotourism from other forms of nature-based tourism. Regulations and restrictions do not necessarily change people's behaviour or attitude towards the environment (Cameron-Smith, 1977). However, this was the strategy being relied upon by the Gosford City Council and New South Wales Fisheries, who were jointly managing the Avoca Beach Rock Platform.

Underlying Tenets of Educating for Ecotourism

The most vexatious question facing environmental educators, and indeed one of the primary objectives of behavioural studies, is to understand the development of environmentally responsible behaviour and to test the predictive power of associated variables. Traditional thinking by environmental education researchers (Ramsey & Rickson, 1977) was generally based on behavioural change models that emphasised linear processes. It was thought an increase in knowledge of ecological systems, environmental problems and issues led to improved awareness and attitudes that subsequently motivated students to act toward the environment in more responsible ways.

The simplicity of this behavioural change model was challenged by Hines *et al.* (1987), who argued that the following variables (in conjunction with knowledge of issues) were associated with responsible environmental behaviour: knowledge of action strategies, locus of control, attitudes, verbal commitment and an individual sense of responsibility. The Hines *et al.* model incorporates more human psychological factors and considers external aspects, so that relationships and synergies exist between the variables themselves. Building on this research, Hungerford and Volk (1990) discovered that the categories of variables probably operate in a linear fashion. However, the variables within each category do not necessarily operate in a linear manner.

Hungerford and Volk's model is based on three main categories of variables, each aiming to predict responsible environmental behaviour:

- entry level variables – those that function as prerequisites to environmentally responsible behaviour and mainly involve the need for *environmental sensitivity*;
- ownership variables – those that make environmental issues very personal through *in-depth knowledge* and understanding; and
- empowerment variables – those that lead to environmental action strategies based on knowledge and skill and *locus of control*.[1]

The variable *environmental sensitivity* is not usually developed in a formal education setting but generally occurs through an individual's contact with pristine environments over long periods of time. It is suggested that instruction techniques need to focus upon ownership of environmental problems and issues. Programmes need to provide skills that empower the individual to investigate environmental issues and be able to take positive environmental action (Hungerford & Volk, 1990).

Cottrell and Graefe (1993) in their first study of boaters on the Chesapeake Bay, Maryland, found a relationship between attitude[2] and pro-environmental behaviour. Their study also found that as self-perceived knowledge of ecology increased, there was a subsequent increase in proactive behaviour. As the level of environmental concern among boaters increased, so did the overall score for General Responsible Environmental Behaviour (GREB).[3]

In a follow-up study, Cottrell and Graefe (1994) targeted knowledge of specific aspects of environmental issues and problems of Chesapeake Bay to ascertain their predictive qualities in determining Specific Responsible Environmental Behaviour (SREB).[4] Knowledge of specific issues had a direct positive effect on responsible environmental behaviour. Cottrell and Graefe's studies (1993, 1994) support prior research of the predictive quality of linear models (e.g. knowledge of specific environmental issues and problems leads to more responsible environmental behaviour). This simple interpretation, however, assumes that respondents will already have developed some pro-environmental attitudes.

In a third study of boaters on the Chesapeake Bay, Cottrell and Graefe (1997: 24) found that 'relationships between various indicators of the three attitudinal components provide a better understanding of behaviour than single-behaviour examinations'. The more complex constructs, such as McGuire's (1969) attitudinal model, provide a better understanding of why people do what they do.

Behavioural research was used by Negra and Manning (1997) to develop principles for designing non-formal park-based environmental education programmes. The authors based their educational programme on the development of Hungerford and Volk's variables in responsible environmental behaviour. They note, 'an environmental education programme that promotes these characteristics may be expected to contribute to fostering environmentally responsible behaviour' (Hungerford & Volk 1990: 12). Negra and Manning's (1997: 10) programmes are based on the assumption that communication is jeopardised by 'imposing inappropriate ideas and expectations that conflict with the listener's values and beliefs'.

In their programme recommendations Negra and Manning (1997) emphasise the need to accommodate and address park visitor subgroups according to characteristics based on their environmental ethics.[5] It is not desirable to direct members or subgroups to specific educational programmes once they enter the park. Members can be expected to be attracted to the subgroup of opportunities that they feel are appropriate for them. Other programming recommendations include the need for information on how humans threaten the integrity of natural systems and the need to emphasise mechanisms to diminish these threats and an outline of the dynamics of environmental problems.

Capacity Building to Enable Communities to Plan for Tourism Using Basic Planning Principles

Generally, ecotourism (Wearing & Neil, 1999) attempts to apply the broad democratic premise that whatever the level of decision-making, ordinary people can be trusted to solve their own problems if they are given the chance. No policy or programme is likely to succeed unless ordinary people are given this opportunity. However, this evolution needs a facilitated process (Wearing & Harris, 1999) that provides the community with the technical knowledge and general directions (Wearing & Mclean, 1997).

For planning systems to be useful, they need to be accessible and easily understood for the local community to assume ownership and control of the process (see Wilson, this volume). One tool that is simplified enough and in frequent use is the Recreation Opportunity Spectrum (ROS). This system provides a conceptually sound planning process that is easily understood by non-experts and can be successfully implemented and monitored by the local community with a minimum of consultative expertise.

The goal of tourism planning is to set in place a strategy that provides a range of recreation opportunities, yet protects the very nature of the area where this recreation takes place. In natural areas, a balance has to be struck between recreational access and conservation. One of the ways in which policy-makers in national parks seek to do this is by designation and zoning systems such as the ROS (Stankey & Wood, 1982). By using the ROS system, it is possible to develop a management plan that will protect sensitive areas, identify settings where visitors may achieve diverse recreation experiences and place management of the process and outcomes in the hands of the local community (Graham *et al.*, 1988).

The basic assumption underlying the ROS is that quality recreation and tourism experiences can best be assured by providing a diversity of opportunities. Stankey and Wood (1982: 7) note: 'A recreation opportunity is defined as a chance for a person to participate in a specific recreational *activity* in a specific *setting* in order to realise a predictable recreational *experience*'.

These ideas about the ROS have been used to develop a formal planning framework called the Recreation Opportunity Planning (ROP) system (Stankey & Wood, 1982). This system is the basic planning framework used by the United States Forest Service (USDA) and the United States Bureau of Land Management.

Using Basic Management Concepts to Enhance Self-Developed Ecotourism Research

To begin the process of devising a management strategy for the Avoca Beach Rock Platform, it was agreed by the Avoca Beach community, local government and the federal government agency, Coastcare, that some basic data on visitors to the rock platform was needed. In addition, in order to facilitate an appropriate education programme, a number of useful visitor characteristics were required. These included:

- Who uses the rock platform and where do they live?
- What activities do visitors typically undertake?
- What are their motivations for visiting the rock platform and how often do they visit?
- What types of management actions and strategies do visitors feel would improve the rock platform?

With this information requirement as a guide, a very basic questionnaire was developed, piloted and administered to 342 visitors to the Avoca Beach Rock Platform on the Easter long weekend of 2000. 'Demographic' questions included age, gender, language spoken at home and suburb of residence. The 'type of visit' questions investigated reasons for visiting the rock platform, how often the platform was visited and the type of activity undertaken. The questionnaire finished by asking respondents what management actions and strategies could be employed to improve the area's environmental integrity and recreational opportunities. (A full breakdown of results from the study can be obtained by contacting the authors.)

Future Strategies for Community Management of the Avoca Beach Rock Platform

The Recreation Opportunity Planning (ROP) system is discussed in the following steps using Stankey and Wood's (1982) introduction to the ROS and using aspects of the Limits of Acceptable Change (LAC) planning process (Stankey *et al.*, 1985; Stankey, 1990). The ROS can be tailored to fit with existing environmental characteristics as evidenced by Jackson's (1987) adaptation of the ROS to an urban setting.

The ROP system should build upon existing systems of tourism and recreation planning, it should be both simple and inexpensive to implement, it should fit with existing planning and management systems used by different organisations and it should ensure that the total range of tourism and recreation opportunities is considered, from highly developed to primitive (Stankey & Wood 1982). These steps are as follows:

Identify area issues and concerns

The process needs to begin with an identification of the community issues that the recreation plan aims to improve and alleviate. Some of these issues had already been identified by the Avoca Beach community and were the motivations behind the initial study to ascertain visitor characteristics. The three main community issues are tourist–resident impacts, environmental degradation of the Avoca Beach Rock Platform and the sense of restriction placed on the host community's access to local recreation sites by the growth of tourist numbers in the area.

Information gathering at community meetings, workshops and through the use of surveys helps the community to make decisions concerning aspects of the planning process. These include recreation demand, the choice of opportunity classes, recreation recommendations and integration with non-recreation resources, carrying capacity (acceptable levels of social and bio-physical impact ascribed to a given recreation site) and an alternative allocation for a recreation opportunity.

One effective way to manage these tasks is through the setting up of a 'steering committee' in this case composed of representatives of the various community organisations, local government officials and Coastcare officers. The Steering Committee is responsible for directing and moulding community information and will eventually assume responsibility for implementation of the plan (Forster, 1994).

Demand and resource capability analysis

Some type of 'recreation demand analysis' must be performed so that proposed patterns of recreation supply can accurately reflect the desired opportunities. For example, the data obtained from this study revealed the residents of nearby Sydney heavily influence recreation demand at Avoca Beach. Demand can also be gauged through the information collected from consultation with the local community.

Opportunity classes then need to be chosen to represent the environmental characteristics of the area. This needs to be a joint process, carried out between the various communities in the local government area and local government itself. The Avoca Beach community would liaise with other community organisations in the area, in order to decide on the most appropriate opportunity classes for the various environments.

Inventorying current opportunities

This stage requires inventorying the present recreation opportunity supply by analysing existing social and managerial settings. From this inventory, the local community can then make decisions about manipulating the social and managerial aspects of a setting to create a desired recreation opportunity. This analysis enables the local community and management to determine if they are providing opportunities consistent with the potential for the setting and whether or not present production meets demand (Stankey & Wood, 1982).

The ROS Management Factors are:

- *access* – including the type of access to and within the area and the type of travel permitted;
- *non-recreational resource uses* – including both the activity and the scale at which it occurs;
- *on-site management or modifications* – including the types of facilities and the extent, apparentness and complexity of the modifications;
- *social interaction* – including the assessment of the type (both resource and social) and amount;
- *acceptability of visitor impacts* – including the assessment of the type (both resource and social) and amount; and
- *acceptable regimentation* – including the nature, extent and level of management control of recreation use.

Tourism and recreation recommendations and integration with other resources

At this stage, decisions need to be made about what combinations of tourism and recreation opportunities are going to be most beneficial to the community. Stankey and Wood (1982) note there are no simple formulae for making these judgements, but that the local community needs to consider information regarding recreation demand, recreation opportunity potential, existing regional supply, the constraints of budget, state government and local government policy, resource capability and the potential use of the same resource base for non-recreation outputs.

The allocation analysis also requires the formulation of a carrying capacity for each recreation site and recreation opportunity. Carrying capacity refers to the level of social and biophysical impact a recreation setting can sustain before it becomes degraded. Stankey (1982) argues that carrying capacity is increasingly becoming a judgemental measure rather then the product of a biologically determined set of conditions. Judgements can vary widely from individual to organisation on what constitutes acceptable levels of carrying capacity.

Action and project plan development

Specifying management strategies through the recreation opportunity class system provides guidance to the local community on activities that are consistent with the objective that has been established through the planning process. The ROS framework makes action and project plans a natural outcome of allocation decisions.

The Recreation Opportunity Planning (ROP) system is a tool that provides the Avoca Beach community with a planning framework. This creates an outdoor recreation system that specifies the recreation opportunities demanded and guides resource inventories to arrive at recreation planning recommendations. It combines recreation opportunity analysis allocation with other resource outputs or impacts of other resource uses on recreation opportunities and ensures consistency between allocation, action and project plans (Stankey & Wood, 1982).

After community and government have set the social and bio-physical limits and indicators for a particular area and opportunity class, community groups can then begin to design a community-directed education programme aimed at maintaining aspects of the site as set down in the Action and Project Plan Development stage of the planning process.

Community-directed Education Programme

Results from the study carried out over the Easter long weekend indicate that the majority of rubbish left on the platform was the result of line fishing. Line fishing requires large amounts of equipment, in the form of fishing line, floaters, sinkers, bait and bait bags. As fishing is a sport that entails many hours spent with a line in the water, fishermen bring food and drink with them to sustain them throughout the day or night. Much of the rubbish left behind, as observed by the researcher and survey volunteers, came from food wrappers and empty drink containers.

In contrast, activities such as exploring, exercising and admiring the scenic beauty require only the natural attributes of the rock platform itself. These types of activities do not generally require large amounts of time spent on the platform and those that undertake these types of activities do not usually bring food and drink with them.[6]

The researchers recorded over 25 languages spoken by line fishermen. More than half of the fishermen in this study spoke a language other than English.

The model proposed by Hungerford and Volk (1990) can be simply applied to the design of a non-formal community-directed education programme on the Avoca Beach Rock Platform. The model contains seven key indicator variables to predict environmental behaviour. Not all of Hungerford and Volk's key indicators can be facilitated through a community-directed education programme. The non-formal environment (as opposed to a classroom) restricts the development of long-term minor and major variables and key indicators of responsible environmental behaviour, such as knowledge of ecology, androgyny, attitudes toward pollution, technology and economics and a personal commitment to issue resolution.

If we accept Hungerford and Volk's assumption that the first key indicator and entry level variable, *environmental sensitivity*, is not usually developed in a formal education setting but through exposure to pristine environments, then many of the programme's participants will have some sort of empathy towards nature due to the many days and hours fishermen spend on rivers and coastlines.[7]

Empathy towards nature, regardless of the participant's environmental ethics, was also found to be a common factor amongst respondents in Negra and Manning's (1997) study. The authors state:

> Responses to environmental behaviour questions suggest that environmental educators may assume that their audience members exhibit some development of several characteristics associated with environmentally responsible behaviour:
>
> 1. Empathy toward the environment.
> 2. Identification with specific environmental issues.
> 3. Awareness of ways to maintain environmental quality.
> 4. A sense of empowerment for achieving desired outcomes. (Negra & Manning, 1997: 17)

We can safely predict that fishermen on the Avoca Beach Rock Platform would have some degree of environmental sensitivity. The degree of *environmental sensitivity* would depend on the homogeneity of Negra and Manning's sample of respondents in this study. Unfortunately, information on respondent's *environmental sensitivity* was not collected in the survey – a limitation of the questionnaire.

The ownership variable, *in-depth knowledge about issues*, is the next key indicator. Cottrell and Graefe's (1993, 1994) findings suggest that knowledge of specific environmental issues result in directly positive and specific responsible environmental behaviour. Development of knowledge should then focus on specific environmental issues facing the Avoca Beach Rock Platform. This would include the building of knowledge related to the collection of intertidal flora and fauna, the overturning of rocks on the rock platform, overfishing and the lethal effect that rubbish left on the platform can have on fish, marine mammals, birds and reptiles (Coastcare, 1998).

Cottrell and Graefe (1997) found no significant relationship between knowledge of general and specific environmental issues and responsible environmental behaviour. Therefore, a community-directed education programme need not educate rock platform visitors on general environmental issues (such as global warming and knowledge of ecological systems) in order for participants to behave responsibly toward specific environmental issues, such as those faced on the Avoca Beach Rock Platform. This is an important point in the design of an education programme, as time and access to tourists is limited.

The next key indicator and empowerment variable, *knowledge of and skill in using environmental action strategies*, has been found to act synergistically with the other major key indicator and ownership variable, *in-depth knowledge about issues* (Hungerford & Volk, 1990). It would stand to reason, then, that both key indicators be taught concurrently with each other. This is useful given that both key indicators can be enhanced simultaneously in a non-formal instructional setting.

Negra and Manning (1997) outline some possible ways that these two key indicators may be simultaneously taught. Anecdotal evidence from the researcher and survey volunteers suggests that many of the fishermen in this study fall under Negra and Manning's ethical subgroup *religiously based anthropocentrism*.[8] The authors note a number of steps when educating this group:

- Emphasise threats to humans posed by environmental degradation and explore the dynamics of specific environmental issues (urine and faeces left on the platform may wash into the ocean and poison the fish stock through the food chain).
- Outline the benefits humans can derive from environmentally responsible behaviour, rather than assuming that maintaining environmental quality for its own sake is important.
- Outline simple environmental action strategies, such as taking your rubbish home when visiting a fishing area. Explain how this simple act can improve the overall quality of the fishing experience (i.e. reduced faunal impact on rock platforms provides more food for fish to feed on: in turn this will attract more fish to the rock platform for fishermen to catch).

The final major variable and key indicator, *locus of control*, 'probably cannot be developed directly in the classroom' (Hungerford & Volk, 1990: 12). However, improved locus of control may result when students have an opportunity to apply learning developed *in using environmental action strategies*. The simple act of taking rubbish home from the rock platform at the end of a day's fishing could be one opportunity that may improve a participant's locus of control.

Placing participants into subgroups based on Negra and Manning's (1997) environmental ethics criteria could not be adapted to an education programme at the Avoca Beach Rock Platform, or in a community setting. State parks in the United States, where Negra and Manning's programmes were trialled, are able to funnel visitors into particular areas (i.e. visitor centres), where programmes can be advertised. State parks also have the facilities and resources to undertake formal environmental education (lecture rooms, interpretation boards, etc.). However, Negra and Manning's research is useful in that it illustrates the need to be sensitive to the tourists' ethics and values towards nature, as outlined earlier.

Recent behavioural research (Cottrell & Graefe, 1997; Gigliotti, 1992; McGuire, 1992) has re-evaluated the importance that attitudes play in the development of responsible environmental behaviour (see also Townsend, this volume). Cottrell and Graefe's (1993) results indicate that attitudes positively influence overt behaviour, that is, an individual's actions toward the environment. The possible design of a non-formal, community-directed education programme presented here assumes that fishermen will have already acquired some values and feeling of concern toward the environment, whether these were developed through formal schooling, public education campaigns or exposure to pristine environments while fishing.

Conclusion

The tenets of ecotourism provide a basis to begin reconciling the issues of natural resource management and tourism that currently challenge many coastal communities in Australia. As a formal strategy, ecotourism seeks to develop a community's capacity to care for and manage their fragile marine environments, while providing a social and economic dividend from the tourism industry. As governments in Australia look to move away from their traditional role of managing natural resources, empowerment of small coastal communities, like Avoca Beach, with the appropriate training and skills will increasingly be seen not only as a more effective way to conserve marine environments but also as obligatory.

The initiation of a community-directed education programme for visitors to the Avoca Beach Rock Platform is a positive first step for the local Avoca Beach community. Through the development of an education programme, a community learns to mobilise its resources, take decisions and enjoy the sense of power and ownership that comes from taking positive action. Carrying out basic research on visitor patterns and characteristics dispels many of the negative myths surrounding tourist visitation and provides data upon which to begin the design and delivery of the education programme. Critical to the long-term sustainability of the Avoca Beach's natural resources and tourism industry is an overall management strategy. We have argued that the Recreation Opportunity Planning system is simple enough for the community to undertake and eventually implement. The result of a planning process and its implementation would begin to cultivate the area's tourism potential and help to create a sustainable industry that will boost the area's economy and provide the community with quality recreation experiences.

Notes

1. The term 'locus of control' refers to 'an individual's belief in being reinforced for a certain behaviour. A person with an internal locus of control expects that he / she will experience success or somehow be reinforced for doing something' (Hungerford & Volk, 1990: 12).
2. Cottrell and Graefe (1993) used McGuire's (1969) attitudinal model to test the relationship between attitude and pro-environmental behaviour. McGuire's (1969) attitudinal model is made up of three components: cognitive, affective and conative. This was operationalised in Cottrell and Graefe's (1993) study by asking respondents to comment on information from the three following areas: cognitive – perceived knowledge of ecology, affective – environmental concern and conative – verbal commitment.
3. General Responsible Environmental Behaviour (GREB) is a construct designed by Cottrell and Graefe (1993, 1994, 1997) to measure a respondent's overall environmental awareness. The general environmental variables used in this study were environmental concern, a verbal commitment to responsible environmental behaviour and a knowledge of ecological systems.
4. Specific Responsible Environmental Behaviour (SREB) is a construct used to measure a

respondent's awareness of specific issues of boating on Chesapeake Bay. These included knowledge of water pollution, dumping on the bay, dumping offshore, enforcement, consequences and commitment to issue resolution.

5. Using cluster analysis, Negra and Manning (1997) grouped their respondents based on the similarity of their responses to environmental-ethics questionnaire items. From this information, the authors devised four potential subgroups in which to tailor park-based, non-formal environmental education programmes. The subgroups are: Spiritually Based Stewardship, Religiously Based Anthropocentrism, Secular Ethical Extensionism and Spiritually Based Biocentrism.
6. This was again observed by the researcher and survey volunteers.
7. Results from the study reveal that line fisherman visit the Avoca Beach Rock Platform, other rock platforms on the Central Coast and other rocky shores of the New South Wales coast on average, once a month.
8. The researchers' anecdotal summation of respondents is supported by Jackson's (1986) study that indicate participants in consumptive activities, such as fishing and hunting, as opposed to appreciative activities such as hiking and birdwatching, are more inclined to view nature from an anthropocentric point of view. Anthropocentric perspectives presume that human moral relationships with nature should be determined solely by human needs.

References

Butler, R.W. (1980) The concept of tourism area cycle of evolution: implications for management of resources. *Canadian Geographer* 24 (1), 5–12.

Cameron-Smith, B. (1977) Educate or regulate? Interpretation in national park management. *Australian Parks and Leisure*, Nov. 34–7.

Ceballos-Lascuraín, H. (1990) Tourism, ecotourism and protected areas. *Seminar Proceedings of the International Union for Conservation of Nature and Natural Resources (ICUN) IVth World Congress on National Parks and Protected Areas, Caracas, Venezuela, 10–12 February* (pp. 84–9). Gland Switzerland: International Union for Conservation of Nature and Natural Resources (ICUN).

Coastcare (1998) *50 Ways to Care for our Coast*. Sydney: NSW Department of Land and Water Conservation.

Cottrell, S.P. and Graefe, A.R. (1993) General responsible environmental behaviour among boaters on the Chesapeake Bay. *Proceedings of the 1993 Northeastern Recreation Research Symposium* (pp. 123–9). Michigan, IL: Park and Recreation Resources Department, Michigan State University.

Cottrell, S.P. and Graefe, A.R. (1994) Specific responsible environmental behaviour among boaters on the Chesapeake Bay: A predictive model. *Proceedings of the 1994 Northeastern Recreation Research Symposium* (pp. 53–6). Michigan, IL: Park and Recreation Resources Department, Michigan State University.

Cottrell, S.P. and Graefe, A.R. (1997) Testing a conceptual framework of responsible environmental behaviour. *Journal of Environmental Education* 29 (1), 17–21.

Evans-Smith, D. (1994) *National Ecotourism Srategy*. Canberra: Commonwealth Department of Tourism, Australian Government Printing Service.

Forster, J. (1994) *Evaluating Community Needs: An Alternative Approach. Municipal Recreation Planning Guide: Sport and Recreation Victoria.* Melbourne: Victorian Government Printing Press.

Fox, A. (1977) *Ranger Uranium Environmental Enquiry*, 2nd Report. Canberra: Australian Government Printing Service.

Gigliotto, L.M. (1992) Environmental attidtudes: 20 years of change. *Journal of Environmental Education* 24 (1), 15–26.

Graham, R., Nilsen, P. and Payne, R.J. (1988) Visitor management in Canadian national parks. *Tourism Management* 14 (1), 44–62.

Heiman, M. (1986) *Coastal Recreation in California: Policy, Management and Access*. Los Angeles: University of California Press.

Hines, J.M., Hungerford, H. and Tomera, A. (1987) Analysis and synthesis of research on responsible environmental behaviour: A meta-analysis. *Journal of Environmental Education* 18 (2), 1–8.

Hungerford, H. and Volk, T. (1990) Changing learner-behaviour through environmental education. *Journal of Environmental Education* 21 (3), 8–21.

Jackson, E.L. (1986) Outdoor recreation participation and attitudes to the environment. *Leisure Studies* 5, 1–23.

Jackson, P. (1987) Adapting the R.O.S. technique to the urban setting. *Australian Parks and Recreation* 3, 26–8.

Jenner, P. and Smith, C. (1991) The tourism industry and the environment. *Condor, The Economist Intelligence Unit Special Report No. 2453*. London: *The Economist*.

McGuire, J.R. (1992) An examination of environmental attitudes among students. Unpublished masters thesis, Pennsylvania State University.

McGuire, W.J. (1969) The nature of attitudes and attitude change. In G. Lindzey and E. Aronson (eds) *The Handbook of Social Psychology*, Vol. 3 (pp. 136–314). Reading, MA: Addison-Wesley.

Negra, C. and Manning, R.E. (1997) Incorporating environmental behaviour, ethics, and values into nonformal environmental education programs. *Journal of Environmental Education* 28 (2), 10–21.

Queensland Department of Tourism, Sport and Youth (1995) *Draft Queensland Ecotourism Plan*. Brisbane: The Queensland Department of Tourism, Sport and Youth.

Ramsey and Rickson (1977) The effects of environmental action and environmental case study instruction on the overt environmental behaviour of eighth-grade students. *Journal of Environmental Education* 13 (1), 24–30.

Stankey, G.H. (1982) Carrying capacity, impact management, and the recreation opportunity spectrum. *Australian Parks and Recreation* 21 (1), 24–30.

Stankey, G.H. (1990) Conservation, recreation and tourism: The good, the bad, and the ugly. In M.L. Miller and J. Auyan (eds) *Proceedings of the 1990 Congress on Coastal and Marine Tourism – A Symposium and Workshop Balancing Conservation and Economic Development*, Newport: National Coastal Resources Research and Development Institute.

Stankey, G.H., Cole, D.N., Lucas, R.C., Petersen, M.E., and Frissell, S.S. (1985) *The Limits of Acceptable Change (LAC) System for Wilderness Planning*. General Technical Report INT–176. Ogden: USDA Forest Service.

Stankey, G.H. and Wood, J. (1982) The recreation opportunity: An introduction. *Australian Parks and Recreation* 21 (1), 6–14.

Wearing, S.L. and Harris M. (1999) An approach to training for indigenous ecotourism development. *World Leisure and Recreation Journal* 41 (4), 9–17.

Wearing S.L. and Mclean J. (1997) *Developing Ecotourism: A Community Based Approach*. Melbourne: Hepper Marriott and Associates.

Wearing S.L. and Neil, J. (1999) *Ecotourism: Impacts, Potential and Possibilities*. Oxford: Butterworth-Heinemann.

Chapter 10
Community Participation in Marine Ecotourism Development in West Clare, Ireland

ZENA HOCTOR

Sustainable tourism, because of its very name, should in theory reflect the principles of sustainable development. The concept of sustainable development is seen to encompass environmental, economic and social issues of development. It is underlain by the principles of holistic planning and strategy making; preserving essential ecological processes; protecting human heritage and biodiversity; developing in such a way that productivity can be sustained over the long term for future generations; and achieving a better balance of fairness and opportunity between nations (Hall & Lew, 1998). Sustainable tourism development should therefore be concerned with its impacts on the environment and sustaining the host community as well as economic growth. The case study detailed in this chapter is concerned with a particular type of sustainable tourism: marine ecotourism. One of the criteria most often agreed as essential to the condition of sustainability in any tourism development is the participation of local people: the host community. Marine ecotourism development has been cited as a solution to problems of regenerating peripheral coastal communities (Lindberg *et al.*, 1996).

Sustainable development stresses education and awareness raising. These enable the individual to make decisions on the best way forward, from an environmental, economic and social viewpoint, in order to benefit both the present and future communities. Sustainable development is to be achieved through cooperative rather than competitive effort and is about moving the balance of power to a more local level. It is also about allowing decisions to be made at 'grassroots level' and enabling the local population to participate in the process.

Participatory development implies the active participation of individuals, groups and communities in shaping their environment and the quality of their living conditions. For the action to be sustained, the process of participation must:

- involve local people in the identification of their own needs and resources and in devising potential solutions;
- enable local people to develop appropriate skills, knowledge and confidence
- empower local people to take initiatives; and
- enable people to participate in the conduct of local affairs.

Participatory development in sustainable tourism projects should result in local ownership of the project. Such ownership will lead to a situation where members of the host community will become increasingly aware of the social and economic benefits that sustainable tourism development can have for their local area. Awareness of the importance of the natural resource to the sustainable tourism product is raised and this results in an increased willingness on the part of the host community to protect the resource and to support local conservation issues. Ultimately, this should lead to the achievement of consensus on how and when local resources should be utilised. Local ownership of the project allows greater control when impacts occur. Impacts are first perceived at the local level and if control is in the hands of local stakeholders remedial action is more immediate. A local sense of ownership and control can engender a long-term commitment to the development and therefore enhance its sustainability. Local participation should result in management strategies being developed at the local level, in conjunction with stakeholders who have a vested interest in the region. Planning and management, therefore, evolve from the bottom up. To allow local participation, control and ultimately ownership of sustainable tourism developments to emerge, the host community must be involved at all levels of planning and decision-making from the outset.

A Case Study of Local Participation in Marine Ecotourism in West Clare, Ireland

The levels of participation that occur in tourism development projects vary. Levels of participation concern the degree to which people are involved or are allowed to have responsibility in the development process. It is important, however, to note that different levels of participation may be appropriate for different situations. In the following case study, the development of marine ecotourism in West Clare, Ireland, is examined in terms of the levels of community participation that have occurred throughout the process, based on comparison with Pretty's typology of participation (Table 10.1).

Situated on the mid-west coast of Ireland at the most western extremity of County Clare, the study area is a peninsula, stretching out to the sea at Loop Head, bounded on one side by the Shannon Estuary and on the other by the Atlantic Ocean. Declining employment in the traditional sectors of fishing and seaside

Table 10.1 Pretty's typology of participation

	Typology	*Characteristic of each type*
1	Manipulative participation	Participation is simply a pretence; there are 'people's representatives' on official boards, but they are unelected and have no power.
2	Passive participation	People participate by being told what has been decided or has already happpened; unilateral announcements are made by project managment without any listening to people's responses; information shared belongs only to external professionals.
3	Participation by consultation	People participate by being consulted or by answering questions; external agents define problems and information-gathering processes, and so control the analysis; the process does not concede any share in decision-making; professionals are under no obligation to account for people's views.
4	Participation for material incentives	People participate by contributing resources (e.g. labour) in return for material incentives; this is commonly called participation, yet people have no stake in prolonging technologies or practices when the incentives end.
5	Functional participation	Participation is seen by external agencies as a means of achieving the project goals, especially reduced costs; people may participate by forming groups to meet project objectives; involvement may be interactive and involve shared decision-making but tends to arise only after major decisions have already been made by external agents; at worst local people may still only be co-opted to serve external goals.
6	Interactive participation	People participate in joint analysis, development of action plans and strengthening of local institutions; participation is seen as a right, not just the means to achieve project goals; the process involves interdisciplinary methodologies that seek multiple perspectives and use systemic and structured learning processes. As groups take control of local decisions and determine how available resources are used, so they have a stake in maintaining structures and practices.
7	Self-mobilisation	People participate by taking intiatives independently of external institutions to change systems; they develop contacts with external institutions for resources and technical advice they need but retain control over resource use; self-mobilisation can spread if government and NGOs provide an enabling framework of support. Self-mobilisation may or may not challenge existing distributions of wealth and power.

Source: Based on Pretty (1995).

tourism, rural depopulation, a community hungry for alternative sources of income, a peripheral coastal location and a diverse range of coastal wildlife habitats and species are the attributes that made West Clare a focus for marine ecotourism development.

The importance of marine based tourism to the the local economy of West Clare has been increasing steadily since the early 1990s. This has mainly been centred around the dolphin-watching industry on the Shannon Estuary at Kilrush and Carrigholt and traditional seaside resort tourism at Kilkee. In 1998, the Irish Marine Institute, Shannon Development Ltd (the Regional Development Authority) and Clare County Council, (the Local Authority) commissioned a pilot study to identify measures for creating a unified focus on marine tourism in the West Clare Peninsula (Marine Institute *et al.*, 1999). The programme of work that was undertaken consisted of an introductory workshop involving local tourism operators, an audit resource survey and SWOT (strengths, weaknesses, opportunities, threats) analysis by commissioned consultants, a second discussion workshop on the survey findings and a final workshop in which working groups, comprised of the local tourism operators, identified specific action recommendations for future development.

The level at which the local community participated in this first study on marine tourism in West Clare was at a consultative level, as defined by Pretty (1995). This level allows people to participate by answering questions through consultation, while the external agents define the problems and information-gathering processes, and so control the analysis. This process does not concede any share in the decision-making at local level in respect of how the study will be undertaken. Although limited from a community participation viewpoint, this first study resulted in a detailed resource audit baseline document, with agreed recommendations to take the process further (Marine Institute *et al.*, 1999).

The change in emphasis to marine 'ecotourism' as opposed to marine 'tourism' in West Clare received a major push when the area was selected as the Irish case study area for the META- (Marine Ecotourism for the Atlantic Area) project. As a partner in this EU Interreg IIc Community Initiative project, the Irish Marine Institute selected West Clare as its project area to allow for implementation of some of the marketing recommendations as identified through the first study carried out in 1999. A full-time project manager was employed by the Marine Institute to implement the project. The main aim of the West Clare META- project was to seek to establish a framework for the development and marketing of a range of integrated marine ecotourism products, aimed at attracting a greater level of participation in marine ecotourism activities, which already existed in the study area (Hoctor, 2001). The objectives were based on assessing tourist motivation for ecotourism, the feasibility of branding a region for ecotourism and developing a package of ecotourism activities.

In the early stages of the project, consultative meetings were held with local statutory agencies and community organisations involved in existing marine tourism activities in West Clare. The purpose of these consultations was to raise awareness of the META-project objectives and to achieve consensus on the most efficient processes to be employed, in ensuring that the objectives of the project were met. The main aim of these meetings was to integrate the META-project with existing tourism developments, and to avoid overlap in work plans and duplication of work already carried out.

The main objectives of the project were subsequently implemented by the project manager, in conjunction with a selected group of marine tourism operators and accommodation providers from the West Clare area. Initially, this group was informed of the anticipated outcomes, their expected role, the role of the META-project manager and the time scale involved in the West Clare META- project. The project manager adopted a leadership role in the early stages of the group's involvement and later, as the project progressed, acted as facilitator to the group in undertaking the following activities:

- development of an ecotourism brand image for West Clare;
- development of a website (www.irrus.com) and marketing literature for the promotion of ecotourism in West Clare;
- development of ecotourism criteria for inclusion under brand marketing, i.e. codes of best practice.

This process ultimately led to the setting up of an independent marketing group by the local tourism operators, which now operates under the ecotourism brand and codes of best practice.

The level of participation of the host community that occurred in the META-project can be compared to that of functional participation, as defined by Pretty (1995). This is the level used by external agencies as a means of achieving project goals. The local community participated by forming a group to meet the project objectives. Involvement was interactive and included shared decision-making. In general, this level of participation tends to arise only after external agents have already made the major decisions. In West Clare, this was the case in terms of the budget, time scale and methodology of the META-project but the basic objectives had arisen from the process of local consultation used through the first study (Marine Institute *et al.*, 1999).

As well as operating at a level of functional participation, the West Clare META-project enabled and empowered the local tourism operators to participate, through provision of technical advice, training programmes and financial support. The META-project manager acted as a facilitator between the local community level and the central agency levels, allowing the bottom-up process to meet the top-down. The process adopted initial leadership through facilitation, which eventually led to the project being wholly owned and controlled at the local level, with the establishment of the independent marketing group operating under the new brand image.

Such participation at a local level is vital to the sustainability of the tourism product being developed. Local ownership of the project from the onset will determine its long-term viability. Continued participation by a core group and perceived success in the branding of the West Clare ecotourism project will encourage participation by others in the future. It is also imperative that local stakeholders can link into the wider regional, national and international planning structures in the future, when projects such as the META- one have been completed. Top-down planning must link with the bottom-up development. Structures to facilitate cooperation and linkages need to be in place to allow for such development. Through the process used in West Clare, linkages were established between the local tourism operators, the Marine Institute, the Regional Development Authority and Local Government Authority to enable the promotion and planning of West Clare as an ecotourism destination, at a national and international level. These linkages now need to be strengthened and developed further, as the META-project has come to an end and the facilitation role of the project manager is no longer available. There is a need to move to the level of self-mobilisation (Pretty, 1995) by the local community in the participatory development process. At this level, the community can participate by taking initiatives independently of external institutions; they can develop contacts with external institutions for the resources and technical advice they need, while retaining control over resource use. Self-mobilisation can spread if governments and NGOs provide an enabling framework of support for the local community. A long-term management plan must be devised by the local group, in consultation with relevant regional and central tourism and funding agents. This will allow the host community to maintain control over the conservation of local resources, while utilising that resource base in a sustainable manner, to enhance local social and economic benefits. Marine ecotourism development is a long-term process that must have the long-term commitment of all the actors involved, i.e. the host community and relevant regional and national agents, to ensure success. All actors must be involved in the planning and management of such development from the outset. The levels of participation undertaken by the various actors may depend on local conditions and planning structures but it is imperative that the host community is included at the higher levels, as defined by Pretty (1995), to ensure the sustainability of the long-term process.

References

Hall, C.M. and Lew, A. (1998) *Sustainable Tourism: A Geographical Perspective*. New York: Addison Wesley Longman.

Hoctor, Z. (2001) Marine ecotourism: A marketing initiative in West Clare. *Marine Resource Series No. 21*. Dublin: Marine Institute.

Lindberg, K. Enriquez, J. and Sproule, K. (1996) Ecotourism questioned: Case studies from Belize. *Annals of Tourism Research* 23 (3), 543–62.

Marine Institute, Shannon Development Ltd and Clare County Council (1999) *Special Interest Marine Tourism in the West Clare Peninsula*. Dublin.

Pretty, J. (1995) The many interpretations of participation. *In Focus* 16, 4–5.

Chapter 11
Marine Ecotourism and Regional Development: A Case Study of the Proposed Marine Park at Malvan, Maharashtra, India

ILIKA CHAKRAVARTY

Examination of the information needs, hopes, concerns and attitudes of the host population in areas of nascent tourism with respect to specific development proposals during the consultation stage has remained an aspect largely neglected in tourism. This chapter discusses resident attitudes towards ecotourism in the context of a proposed marine park in Malvan, Maharashtra, India. Following a brief introduction on the marine environment in the proposed marine park area, the study discusses resident perceptions and opinions regarding the proposed park. It also identifies the concerned population groups within the community in the proposed park area. Discussion then follows on the information needs of residents in the proposed park area, with special reference to the availability of information and residents' familiarity with the proposed park development plan. Lastly, suggestions are outlined for the proposed park, for the tourists and for the residents around the proposed park area based on the findings of a field survey.

A Profile of the Marine Environment in the Proposed Marine Park

The marine environment is under increasing pressure with more people using the sea for food, income and recreation. Waste and run-off from the land, various forms of pollution, oil spillage, over-fishing, beach litter and so on, have contributed to the degradation and loss of marine ecosystems. Malvan is one of the few remaining places untouched by human interference, where the luxurious growth of flora and fauna exist undisturbed in pristine glory. Several species of sea grasses, algae (including the rare green algae) and mangroves, including *Apiculata, Kandelia candel, Sonneratia alba, S. aseolaris, A. officialis, Agadocha,* exist in abundance. The coast is also home to various species of marine animals, including sponges, soft and hard corals, sea anemones, stomato pods, molluscs, sea cucumbers, sea urchins, star fishes and oysters (NIO, 1980). The Malvan coast, therefore, constitutes an important place in the nation's natural heritage and provides opportunities for recreation, education and research.

The Malvan coastline is narrow and dissected, with the transverse ridges of the Ghats extending as promontories in the Arabian Sea. The shore is characterised by sandy stretches, interspersed by rocky cliffs, embankments, submerged shoals and chains of exposed offshore islands as the Vengurla Rocks, south of the Sindhudurg Fort. The surface water temperatures show erratic fluctuations and range from 25–31°C. The surface salinity remains constant at 35% from October to May and reduces to 20% during the monsoons (June to September). The dissolved oxygen levels in the seawater, which play an important role in controlling the distribution of fish, show large variations in time and space, ranging between 3.7 and 5.2 millilitres. The transparent waters make objects clearly visible within a depth of 1.5–2.0m after the monsoon. Most of the marine flora and fauna from the inter-tidal area and coral reefs are exposed during low tide (NIO, 1980). The bay has narrow openings through which the tide flushes water in and out, bringing nutrients and oxygen to its diverse marine inhabitants. Boating and watching the behaviour of marine animals aboard licensed vessels and enjoying the hospitality of warm and friendly local fisherman are added attractions.

The Marine Park Proposal and Ecotourism

Typically, a marine park is an area of sea specially dedicated to the protection and conservation of biodiversity, particularly endangered species of marine flora and fauna, their habitats and associated cultural resources (such as shipwrecks and archaeological sites), for present and future generations (NIO, 1980). Protective measures are generally taken to ensure conservation while allowing maximum compatible usage, particularly by ecotourists, aimed at generating development in an environmentally responsible manner. The National Institute of Oceanography (NIO), Goa, a premier research institution of the Council of Scientific and Industrial Research (CSIR) Government of India (GoI), is engaged in marine biology, coastal and high sea oceanography and conservation of the marine resources along the Indian coast. With support from the various Departments of the Government of Maharashtra (GoM), i.e. the Science and Technology (S&T) Cell, the Department of Tourism (DoT), the Malvan Port Authority and the World Wildlife Fund (WWF), Mumbai, the NIO came up with a proposal for the development of a marine park at Malvan. The proposal was submitted by the NIO to a High Level Coordination Committee in October 1979, which discussed various issues concerning the marine park. The preparation of an action plan followed, with approval from the GoM (NIO, 1980). The Malvan Municipality and the Maharashtra Tourism Development Corporation (MTDC) became the designated agencies responsible for the administration of the proposed park.

For the conservation of marine life and the effective implementation of regulations, the identification of the proposed marine park involved demarcation through a 'two-zone system'. The inner area represented the 'core zone' while the outer area represented the 'buffer zone'. The core zone covers the area of the Sindhudurg Fort, the Padamgad Island and the surrounding submerged and exposed rocky structures. Exploitation of the marine resources and disturbance of the natural environment is strictly prohibited in the core zone, which requires the strictest conservation measures (NIO, 1980). The buffer zone covers the entire Malvan coast and its fishing villages. The northeastern border of the buffer zone runs near the shoreline, approximately 50m from the shore, such that the buffer zone does not include Malvan town. In the southeastern side, it covers the semi-circular sandy beach about 500m parallel to the shore. Within the buffer zone, activities like dredging, effluent discharge or any form of damage to the marine environment (except fishing operations of a traditional nature) have been prohibited so as to prevent adverse effects on the marine environment in the core zone. The offshore tracts being one of the most valuable areas for fishing activities, fishing by poison, shipping or using dynamite or any type of explosives have been prohibited within the zone. The development of Malvan town has also been conceived in such a manner that it has the least possible effect on the buffer zone (NIO, 1980). That is to say, although the buffer zone encompasses a sizable portion of the five fishing villages on the Malvan coast, the 1980 NIO Report states that the development of the marine park should in no way serve to dislocate the resident population along the coast.

The recent thrust of the Maharashtra Tourism Development Corporation (MTDC) has been the promotion of ecotourism through environmentally sustainable efforts. In view of this, the development of the marine park has been conceived as a means for preserving the marine environment as well as promoting ecotourism through a range of coastal activities, such as a marine aquarium, a dolphinarium, a marine museum and a nature interpretation centre. Plans also exist for developing a scuba diving centre for those interested in diving excursions to coral reefs off the Malvan coast. The centre will be run by professional diving instructors and will also provide certificate courses in scuba diving by involving participation from the Professional Association of Diving Instructors (PADI) (MTDC, 1998). According to the MTDC, the proposed location of the marine park in proximity to the Sindhudurg Fort and *en route* to the major tourist destinations of Mumbai and Goa holds good prospects for generating income and employment for the coastal communities in its vicinity. According to the NIO feasibility report, the approximate cost for the development of the marine park (for the first five years) has been estimated at Rs. 25 lakhs (£2.5 million) at 1980 prices (NIO, 1980). Details have not been furnished on the sources of funding and the costs are to be divided among the central and state governments, the Malvan Municipality and the private sector.

Study Area and Background

The coastal stretches adjoining the proposed marine park area are highly populated, with nearly 3000 fisher folk residing in more than 800 houses along the Malvan coast. The local fish workers presently use the area for activities including fishing, the drying of fish, walking and swimming. The fish workers have also accumulated a vast knowledge base and a variety of technologies tailored to the specific ecological niche of the region. The indigenous fishing communities of Malvan practise the traditional sea courts, where the community heads assemble at the place of worship to hear and decide on issues within the community.

With the state-sponsored agents advocating intensive development through tourism by trying to capture the ruling market by supporting the marine park at the cost of the local communities dependent on the common property resources (CPRs), the issues of sustainable use and control over scarce natural resources have been a locus of confrontation between the traditional fish workers and farmers. This has resulted in the confrontation of often antagonistic and conflicting systems of rules and values related to the use of scarce natural resources. The close and peaceful coexistence of the past with regard to the traditional resource users and the environment has been imperiled considerably. Unfortunately, Indian law does not recognise socially established community rights that are basic to the livelihood of the communities dependent on the CPR. In view of this, the underlying principle that has been in vogue is that any land / area without an individual *'patta'* (document supporting ownership) is *terra nullius* (nobody's land) and thus open to acquisition by any party. Traditional skills and knowledge (unlike modern technical skills) do not appear to be an imperative for national development. As such, the livelihoods of the local people, their culture and their ecological space have all been seriously affected (Roy Choudhury, 2000). To the residents of the proposed marine park area, the sea and the land have been the basis of economic sustenance, culture and identity of the community. The marine park could lead not only to nearly complete exclusion of the local population from vital resources but also displacement of their basic rights in pursuing their traditional livelihood.

The displacement of the fish workers within the proposed marine park area, in the absence of any other means of alternative employment and for those who have traditionally practiced fishing, does not seem justifiable. Moreover, rehabilitation would also be a difficult task, as there are no vacant lands available and the fish workers may not be willing to move to other areas once dispossessed of their ancestral lands. According to Ramesh Dhuri, Executive Member of the National Fishworkers' Forum, (Park) the demands of the *Malvan Machchimari Sangh* (Malvan Fishermen's Union) are as follows:

(1) assurance from the government that the boundaries of the marine park should not be extended beyond Malvan, Vengurla and Deogad *taluka*'s (presently under the jurisdiction of the proposed project);
(2) assurance from the government that no fisherman will be evacuated from his lands for the construction of the marine park;
(3) arrangements made to protect the seawater and marine wealth through the development of a marine estate; and
(4) the establishment of an advanced fishing centre with laboratory facilities.

Apart from the fishing industry, there are agriculturists cultivating paddy, coconut, *kokam* and various inferior crops within the proposed park area. If the tourism project materialises, more than 250 acres of thickly populated, fertile lands will be under the threat of acquisition. Moreover, coconut cultivation, which has been inherited for generations and bestows employment for many, would be stopped forever. Besides, various religious (temples, mosques, churches) and public institutions (schools, *madras*'s, cremation grounds) are also located in the same area. These would be completely submerged and this would hurt the feelings of the local people. However, according to Shri Dyanesh Dewoolkar, President of the *Malvan Machchimari Sangh,* in the maps of the NIO Master Plan the area has been represented as uninhabited tracts with sandy bars.

The *Machchimari Sangh* has been accused by the Government of an orthodox approach, being confined to the protection of the interests of a single community. Environmental and resistance groups have also protested against the marine park and seminars have been organised highlighting the environmental problems of the project. They have adopted the issue of land being sold off in clear violation of the coastal regulation zone (CRZ) covering residential areas, in other words, the issue of the GoM flouting its own laws.

Residents' Perceptions and Opinions on the Potential Impacts of the Proposed Marine Park

The objectives of the study on which this chapter is based were as follows:

(1) to investigate the residents' attitudes regarding the impacts of the proposed marine park;
(2) to identify resident opinions about the marine park and the community groups most likely to be affected by the project;
(3) to assess the information needs of the residents surrounding the proposed marine park area; and
(4) to examine the familiarity of the residents with the park development plan and the influence of such exposure on residents' perception and general opinion on the marine park.

The likely impacts of tourism from the proposed marine park were determined by identifying residents' perceptions of the potential benefits and costs of the project for the community. To identify interest groups within the community, differences between residents' socio-demographic subgroups were examined with respect to their perceptions of the potential impacts and general opinions on the park. The adequacy of information possessed by the residents on the park development plan was examined following a qualitative assessment of the information contained in the NIO Master Plan and an evaluation of residents' familiarity with this.

Sample description and methodology

After a review of the pertinent literature and consultation with local planning and municipal authorities, the following categories were identified as being those which would be expected to enable the researcher to obtain the most relevant information:

(1) five coastal villages – Kinara Bundar, Wairy Bandh, Wairy Bhutnath, Tarkarli and Devbaugh – identified with respect to distances from the proposed park area, i.e. 2 km, 4 km, 6 km, 8 km and 10 km respectively.
(2) diverse categories of occupation, i.e. fish workers/farmers/labourers, retail/services, professional/executive/entrepreneur, students, retired and unemployed;
(3) ownership of land (landowners in designated park area and others);
(4) use of the park area (users and others).

Table 11.1 presents a profile of the five fishing villages in the proposed marine park area.

An interview schedule was designed for obtaining information from a sample of the resident population inhabiting the five fishing villages. The interviews, extending from 4–18 February 2001, were conducted in an informal atmosphere wherein the respondents were encouraged to express their views freely on the proposed project. Open-ended questions ensured flexibility and stimulated conversation from the residents on the possible benefits and costs of the project. This ensured that the ideas came first from the respondents and impacts that were not considered important *a priori* were not overlooked. The latter half of the interview, which was structured with closed questions, helped in eliciting opinion on specific

Table 11.1 Profile of the fishing villages in the proposed marine park

Name of Village	*Population*	*No. of households*	*No. of households interviewed*
Kinara Bundar	590	154	112 (72.73%)
Wairy Bandh	643	203	156 (76.90%)
Wairy Bhutnath	729	107	98 (91.60%)
Tarkarli	1779	408	149 (36.50%)
Devbaugh	2944	608	223 (36.70%)
Total	6685	1480	738 (49.86%)

Note: Figures in parentheses indicate percentages of the households interviewed to the total.
Source: Census of India (1991).

potential impacts, unless these had already been mentioned in response to the open questions. An adult member of every second household was interviewed. In the case of refusals, which were rare, the adjacent householder was interviewed instead.

The survey resulted in 738 completed interview schedules, i.e. 50% of the total households (1480) and 11% of the total population of the five villages (6685). Comparison of the sample profile and the actual demographic characteristics of the village population with the census figures revealed that the sample was slightly skewed to the older groups, males and landowners. Residents' general attitudes concerning the positive and negative impacts of the proposed park were examined by ranking their opinions on a seven-point Likert scale. In order to identify population groups within the community who were most likely to be affected by the proposed park, chi-square analysis was used to examine the relationships, if any, between residents' socio-demographic profiles, their perceptions of the potential impact of the park (determined by responses to closed questions), and their general opinion on the project. These included, for example, positive impacts – opportunities for cultural exchanges, a more attractive park areas, more attractive village, new and improved services, improved infrastructure, etc. – and negative impacts – congestion of services and stores, congestion of beach facilities, inflation, spoiling of natural environment, problems of land expropriation, etc. The activities and role of the *Malvan Machchimari Sangh* in respect of the proposed marine park was also considered in detail. Meetings and discussions were held with the fisher workers and leaders of the *Machchimari Sangh* as well as with the officials of the Malvan Municipality.

Empirical Results

Residents' perceptions of the potential positive and negative impacts of the proposed park were one of the most important criteria to be taken into consideration. These were examined with reference to two broad categories: visitor presence and the development of tourism infrastructure. Residents' perceptions of the potential positive impacts of the marine park relate to economic benefits – i.e. the creation of jobs and the generation of income – which were cited by 83% and 61% of the population respectively (Table 11.2). As regards socio-cultural impacts, opportunities for social interaction and cultural exchange as an outcome of visitor presence were cited by 26% and 9% of the sample respectively. Improvements in infrastructure and services were mentioned by just over half the respondents. It is interesting to note that these aspects were rarely mentioned in response to the open-ended question (mostly less than 10% for all, except for 11% for new or improved services), suggesting that they were perhaps less readily thought of as benefits.

The most serious negative concerns of residents related to visitor presence was the possible increase in traffic and noise levels, respectively, as cited by 61% and 35% of the sample in the closed questions. As regards tourism infrastructure, 65% and 49% of the respondents supported the raising of local taxes and the restriction of the rights to the use of the park area (Table 11.3). Approximately half the respondents

Table 11.2 Residents' perceptions of potential positive impacts of the proposed marine park

Positive impacts of the proposed marine park	*Percentage of respondents mentioning impacts (open questions)* (n = *738)*	*Percentage of respondents mentioning impacts (closed questions)* (n = *738)*
Related to visitor presence		
1. Creation of jobs	39	83
2. Increased income	26	61
3. Increased opportunities for social interaction	9	26
4. Increased opportunities for cultural exchange	4	9
Related to tourism infrastructure		
5. Improved infrastructure, roads, parking, etc.	4	53
6. New or improved services	11	52
7. More attractive park areas	8	28
8. Improved recreation facilities	4	27
9. Conservation of park area	7	24
10. More attractive village	4	19

Source: Based on the response to the survey of the residents conducted in Kinara Bundar, Wairy Bandh, Wairy Bhutnath, Tarkarli and Devbaugh during 4–18 February 2001.

mentioned that they regularly used the proposed park area for residential purposes, fishing and recreational activities. Comments on the possible restrictions of residents' rights indicated that many were concerned about payment for use of an area to which access had been previously free. Many local fish workers expressed the belief that they would not have the same access to the quay as before.

In response to the closed questions, 36% of the respondents revealed their concern about possible disputes arising over land acquisition. Comments concerning the possibility of land disputes suggested that 18% of the respondents feared that landowners in the designated area would not be adequately compensated and that land would be subject to expropriation. Many respondents cited examples of land acquisition-related problems, for example in relation to the beaches of Shiroda, Mochemmad and Mithbhav, associated with the construction of luxury resorts by hotel groups. With most of the land being privately owned residential areas, the Malvan Municipal authorities would have to negotiate with the landowners for acquiring this.

Many respondents also felt the park authorities were being too ambitious with the proposal for the marine park. Instead, they advocated the development of tourist amenities such as the provision of improved parking facilities at the beach. These would, in the respondents' opinion, generate revenues from the tourists and help reduce the risks of incurring debts though park operation costs and of increasing taxes. A related concern was of the sizable expenditures jeopardising other

Table 11.3 Residents' perceptions to potential negative impacts of the proposed marine park

Negative impacts of the proposed marine park	*Percentage of respondents mentioning impacts (open questions)* (n = *738)*	*Percentage of respondents mentioning impacts (closed questions)* (n = *738)*
Related to visitor presence		
1. Increase in traffic	15	61
2. Increase in noise	8	35
3. Inflation	4	30
4. Change in village character	9	22
5. Congestion of services and stores	5	23
6. Congestion of beach facilities	5	6
Related to tourism infrastructure		
7. Increase in local taxes	11	65
8. Restriction of residents' rights regarding use of park zone	17	49
9. Disputes over land acquisition	9	36
10. Inflation of land prices	3	24
11. Problems of land expropriation	2	18
12. Spoiling of natural environment	2	12

Source: Based on the response to the survey of the residents conducted in Kinara Bundar, Wairy Bandh, Wairy Bhutnath, Tarkarli and Devbaugh during 4 February 2001 to 18 February 2001.

community projects. Many respondents suggested ways of spending the money: for example by upgrading the village roads or the sewage system, building a swimming pool, and so on. The differences observed in responses to the open-ended and closed questions (Tables 11.2 and 11.3) suggests that many residents may well have little notion about the possible consequences of the marine park, whether positive and negative.

Residents' general opinions on the marine park when ranked on the seven-point Likert scale enabled the researcher to judge how favourable each item was in respect to the construct of interest. The Likert scale permitted the computing of two measures of response: the percentage of the respondents agreeing/disagreeing to the statement, and a mean score reflecting the intensity of the agreement. As seen in Table 11.4, despite the numerous concerns of residents with respect to the proposed park, more residents were strongly in favour of the project (19.5%) than strongly opposed to it (14.6%), which suggests that for residents the perceived advantages compensated for the perceived disadvantages. It should be noted that a little over a quarter of the respondents (26.8%) were neither for nor against the park.

Table 11.4 Residents' general opinion concerning the proposed marine park

Support for the proposed park	*Percentage*
1. Strongly opposed	14.63
2. Moderately opposed	6.91
3. Slightly opposed	4.88
4. Neither opposed nor in favour	26.83
5. Slightly in favour	12.61
6. Moderately in favour	14.63
7.Strongly in favour	19.51
Mean score	4.37

Note: Respondents were asked to indicate their position on a seven-point scale given that one was strongly opposed and seven was strongly in favour and four was neither opposed nor in favour. The other scales reflected intermediate categories.
Source: Based on the response to the survey of the residents conducted in Kinara Bundar, Wairy Bandh, Wairy Bhutnath, Tarkarli and Devbaugh during 4–18 February 2001.

Concerned Population Groups within the Community in the Proposed Marine Park Area

In order to identify population groups within the community that were most likely to be affected by the proposed park, a chi-square analysis was used to examine the relationships, if any, between residents' socio-demographic profiles, residents' perceptions of the potential impacts of the park and their general opinion on the project. The hypotheses were as follows:

- residents' perceptions about positive/negative impacts depend on residential location;
- residents' perceptions about positive/negative impacts depend on occupation;
- residents' perceptions about positive/negative impacts depend on ownership of land in park area;
- residents' perceptions about positive/negative impacts depend on use of the park area.

The results of the differences in resident perception and opinion across these socio-demographic categories have been summarised in Table 11.5.

By socio-demographic subgroups, the strongest relationships were found to exist between residential location, occupation and use of park area. The residents living closer to the park perceived more potential negative impacts, as did the commercial fish workers. Among the fish workers, 86% cited that the restriction of residents' rights in the park area would be a possible disadvantage, compared with 49% for the total sample. Residents who already used the park area tended to perceive more potential impacts, both positive and negative. A possible explanation for this relationship is that they were more interested in the activities taking

Table 11.5 Association between resident perception and opinion to the proposed park across selected socio-demographic characteristics (number of observations 738)

Perceptions	*Resid-ential locations*	*Occup-ation*	*Owner-ship of land in park area*	*Use of park area*	*Supplementary observations*
Positive impacts					
1. Creation of jobs	N.S.	N.S.	0.05	N.S.	Landowners mention less often
2. Increased income	N.S.	N.S.	N.S.	0.05	Users mention less often
3. Opportunities for social encounters	N.S.	0.01	N.S.	0.05	Students and users mention less often
4.Improved recreation facilities	0.01	0.05	N.S.	N.S.	Residents living further from the park and students mention more often
5. Conservation of park areas	N.S.	N.S.	N.S.	0.05	Users mention more often
Negative impacts					
6. Increase in noise	0.001	N.S.	N.S.	N.S.	Residents living closest to park mention more often
7. Change in village character	0.05	N.S.	N.S.	0.05	Residents living closest to park and users mention more often
8. Disputes over land	0.01	0.05	N.S.	N.S.	Residents living closest to park mention most often: no clear pattern for occupation
9. Restriction of residents' rights	N.S.	0.05	N.S.	N.S.	Fish workers mention more often
10. Inflation of land prices	N.S.	N.S.	0.05	N.S.	Landowners mention less often
11. Increase in local taxes	N.S.	0.05	N.S.	N.S.	Fish workers and farmers mention more often; students and retired people less often
12. General opinion on the project	0.01	0.01	N.S.	0.01	Residents living closest to park and fish workers have more negative opinion; users are less indifferent than others

Note: N.S. denotes non-significance of the chi-square statistic at the 5% level of significance, while 0.01 and 0.05 denote the 1% and 5% levels of significance respectively.
Source: Based on the response to the survey of the residents conducted in Kinara Bundar, Wairy Bandh, Wairy Bhutnath, Tarkarli and Devbaugh during 4–18 February 2001.

place within the park and had consequently thought more about the project and identified both potential advantages and disadvantages.

Differences between the residents' socio-demographic subgroups with respect to perceived impacts would appear to be limited and rather inconsistent with observations made in established tourist communities more generally. The lack of significant associations, however, is more revealing; for it is interesting to note that very few relationships were observed for the residents owning land within the designated park area. Since landowners would clearly be affected most by the creation of the park, it was believed that their perceptions of the positive and negative impacts would differ markedly from other respondents. However, this did not appear to be so. Likewise, it was believed that business people would benefit most from the economic impacts of the project and perceive more potential advantages. This too was not apparent from the findings.

Analysis of the relationships between socio-demographic characteristics and general opinions tended to support these findings. Thus, analysis of differences in opinions according to residential location showed that those residents living closest to the park held the least favourable opinions, while analysis of differences across occupational categories showed commercial fishermen to be least in favour of the project. Thus, on a seven-point scale, commercial fishermen had mean scores of 3.5 and 3.2, respectively, compared to 4.4 for the entire population. Opinions also differed significantly with respect to residents' use of the park area. Users were found to hold stronger opinions, either for or against the project, than the non-users. As suggested previously, this may be explained by the fact that they, as users, were more concerned about the future of the area and therefore less likely to be undecided or indifferent. There also existed landowners who had not yet made a firm decision. From comments received during the interviews, it seemed that their indecision flowed from the lack of clarity regarding land acquisition and, more particularly, from compensation payments.

Resident Information Needs in the Proposed Marine Park

Community involvement in tourism planning can assume various forms and serve several purposes. The basic aim of community participation programmes should be to provide concerned citizens with adequate information, such that they can play a meaningful role in the management of tourism-related resources. According to Lucas (1978: 52): 'If full information was not available on issues under consideration, opportunities, or even rights to participate become meaningless'. Identification of the information needs of residents in the proposed marine park therefore assumes considerable importance. To gain insights into resident information needs, an attempt was made to determine:

- what information was available on the issue at stake;
- familiarity of the residents with the information available; and
- whether any relationship exists between familiarity, perceptions of potential impacts and the general opinion on the project.

Issues and availability of information

The major issues were identified as being those potential impacts perceived by more than 30% of the residents. Thus, on the positive side, the likely impacts of the marine park through the creation of jobs and incomes, services and improved infrastructure turned out to be important. On the negative side, increasing traffic and noise, increasing local taxes, the restriction of residents' rights and disputes over land acquisitions were identified. In respect of the potential positive aspect of the proposed park, it was not surprising to find that the NIO study coming out strongly in favour, with details on the potential of the park for creating jobs, generating income and providing improved infrastructure and services.

Information on the potential negative aspects of the proposed project appeared less adequate in the feasibility study. For example, according to the park development proposal the increase in traffic and noise from the project would depend largely on the number of tourists attracted to the park and could, in the long term, be unavoidable. Certainly, predictions suggest that traffic levels will increase and the feasibility study points to some of the potential traffic to the park already passing through the fishing villages *en route* to tourist destinations such as Goa. The NIO study has little information to allay concern over the impact of local taxes. Such concerns arise from the fear that tax increases might be needed to conserve park operations, given that the park facilities would be administered by a state government agency, the Malvan Municipality. The feasibility study did not identify any possible sources of financing for either the capital or operating costs and residents have not received any indication of the likelihood or degree of rise in local taxes. A related concern, voiced by a number of residents in response to the open question on the possible negative impacts of the park, is that other community projects would have to be forgone.

With respect to the restriction of residents' rights, the feasibility study contained information on the facilities that would have to be provided. Although information exists on the development proposed for the wharf areas, the implications of such development, particularly on the local fish workers, are not clear. This appears to have led them to believe their rights would be restricted. Access to tourist attractions (that is, the marine aquarium, dolphinarium, marine museum and nature interpretation centre) would increase the overall attractiveness of the park to the visitors. However, little consideration appears to have been given to the consequences of increased activity, particularly on fishing operations. Moreover, the park feasibility study appears to have left the Malvan Municipal authorities to deal with the thorniest problem of land acquisition. Consequently, there is some concern from both landowners and other residents and the former might not be adequately compensated. The concern is heightened by the fact that the source of financing for land acquisition has not yet been determined.

In view of this, it appears that although the feasibility study contains information on the likely impacts of the park, residents' concerns have not been covered adequately. More information is required on the expected impact of the park on local taxes, rights of the residents (particularly fish workers) and on the compensation of landowners. Comments received during interviews confirm that resident opinion could have been more favourable, had they been better informed on these issues.

Residents' familiarity with the proposed marine park development plan

Slightly over half the respondents were vaguely familiar with the area proposed for the development of the marine park. The majority of the residents knew little about the amenities and activities proposed for the park, the financing of the project or its likely impacts on the local economy. This became evident from the empirical investigation reported in Table 11.6. In the questions pertaining to the proposed amenities and activities, only one aspect (rental cottages) was mentioned by more than 30% of those interviewed. It is perhaps fair to point out that some of the amenities planned for the park, such as the marine aquarium, the dolphinarium and the nature interpretation centre, were more difficult to conceptualise and, as a result, probably less easily remembered. The financial and administrative roles of the Municipality were also not very clear to most respondents who had little idea about the costs of the proposed project and sources of funding.

The general unfamiliarity of residents with the park development plan is perhaps surprising given that the plan model was displayed at various social gatherings,

Table 11.6 Residents' familiarity with the proposed park development plan

Questions concerning proposed development	*Percentage of correct replies*[a] (n = *738)*	*Correct answers*
1. Which area is proposed for development?	54	Coastal area between the Sindhudurg Fort and Malvan town encompassing five coastal villages
2. What activities and amenities are planned for the park?	No aspect mentioned by more than 50%	Aquarium, dolphinarium, marine museum, nature interpretation centre
3. How many vehicle entrances are planned for the park?	27	2 entrances
4. How much is the projected total cost of the project?	11	Rs. 25 lakhs (1980 prices)
5. What are the proposed sources of funding	9	NIO report does not mention of any
6. Would you be responsible for park administration development?	29	NIO report does not mention
7. How much is the park expected to generate in terms of income and employment?	Less than 5%	NIO report does not mention any figure

[a] In the case of questions requiring quantitative replies (questions 3, 4, 6 and 8), answers within 10% of the exact figure were considered correct.
Source: Based on the response to the survey of the residents conducted in Kinara Bundar, Wairy Bandh, Wairy Bhutnath, Tarkarli and Devbaugh 4–18 February 2001.

such as the *Paryatan Mahotsav* (Tourism Festival) which took place on 26 January 1992 in Malvan. Moreover, the project had also generated considerable coverage in the local press in the light of the ongoing controversy between the government and the local fishing community under the *Malvan Machchimari Sangh*. Residents living closest to the park entrances, the landowners in the proposed park zone and users of the park area, who were most likely to be affected by the project, were found to be the most familiar with the development proposals. Not surprisingly, at Devbaugh, being furthest away from the proposed park area many residents felt that the park would have little or no effect on their lives.

According to Sadler (1979), in environmental affairs non-participation is the norm of the vast majority of citizens. Many citizens trust their selected representatives to act in their best interests, while others are simply not sufficiently interested to seek information (Keogh, 1990). This could have been one of the factors influencing the lack of awareness about the marine park. Yet there were many others who were eager to know about the proposed project, even though all they had received from the project authorities were some vague statements. Another factor contributing to the general lack of familiarity with the development plans was the fact that the onus was largely on the villagers to obtain information on the project by visiting the Municipality or the office of the *Machchimari Sangh*, where they could examine the feasibility report and other plans. However, it is not sufficient simply to make information available; it must also be in a form that is readily understood.

Residents' Information, Perceptions and Opinions of the Proposed Marine Park

It was felt necessary to understand whether familiarity with development proposals might be a factor in influencing resident perceptions and opinions. Given the large number of questions relating to the amenities and activities planned for the marine park, a single index was calculated to reflect the overall familiarity of each resident with the physical aspects of the proposed project. In calculating the index, equal weight was given to each of the following items mentioned by respondents: provision of rental cottages, the aquarium, the dolphinarium, the marine museum, the nature interpretation centre. Other information variables reflected residents' awareness of the financial and administrative aspects of the proposed development plan.

The chi-square analysis revealed that most residents who were more familiar with the development plan anticipated a greater number of positive impacts than others. Thus, residents most familiar with the activities and amenities of the proposed park mentioned the following advantages more frequently: the creation of jobs, increased incomes, improved infrastructure (parking, roads), new or improved services, more attractive park areas, improved recreation facilities, conservation of the park area (Table 11.7). However, the findings were not statistically significant. Nonetheless, the analysis also revealed that residents who were familiar with plans also tended to perceive potential disadvantages, such as increases in

Table 11.7 Residents' perceptions of potential impacts according to level of familiarity with proposed park development plan

Perceptions	*Chi-square analysis to identify differences according to familiarity with proposed park amenities and activities*[a] (n = *738)*	
	Level of significance	*Relationships observed*
Positive impacts		
1. Creation of jobs	0.05	Residents more familiar mention more often
2. Increases income	0.01	Residents more familiar mention more often
3. Increased opportunities for social encounters	N.S.	
4. For cultural exchange	N.S.	
5. Improved infrastructure	0.01	Residents more familiar mention more often
6. New or improved services	0.01	Residents more familiar mention more often
7. More attractive park area	0.01	Residents more familiar mention more often
8. Improved recreation facilities	0.01	Residents more familiar mention more often
9. Conservation of park area	0.01	Residents more familiar mention more often
10. More attractive village	N.S.	
Negative impacts		
11. Increase in traffic	0.01	Residents more familiar mention more often
12. Increase in noise	0.01	Residents more familiar mention more often
13. Inflation	N.S.	
14. Change in village character	0.01	Residents more familiar mention more often
15. Congestion of services, stores	N.S.	
16. Congestion of beach facilities	N.S.	
17. Increase in local taxes	0.05	Residents more familiar mention more often
18. Restriction of residents' rights	0.01	Residents more familiar mention more often
19. Disputes over land	0.05	Residents more familiar mention more often
20. Inflation of land prices	N.S.	
21. Land expropriation problems	N.S.	
22. Spoiling of natural environment	N.S.	

[a] Determined on basis of residents' awareness of following amenities and activities planned for the park: aquarium, dolphinarium, marine museum, nature interpretation centre, rental cottages, boat and sail boat facilities, etc. N.S.– Not significant at 5% level.
Source: Based on the response to the survey of the residents conducted in Kinara Bundar, Wairy Bandh, Wairy Bhutnath, Tarkarli and Devbaugh during 4–18 February 2001.

Table 11.8 Residents' general opinion according to familiarity with proposed park development plan

General opinion	*Percentage distribution (better-informed)*[a]	*Percentage distribution (less-informed)*[a]
1. Strongly opposed	11.70	15.60
2. and 3[b] (moderately opposed and slightly opposed)	6.70	13.40
4. Neither opposed nor in favour	16.60	30.20
5. Slightly in favour	15.00	11.80
6.Moderately in favour	20.00	12.90
7. Strongly in favour	30.00	16.10
Total	100.00	100.00
Mean opinion score	5.00	4.20

Chi-square calculated from frequencies=11.70 (significant at 5% level).
[a] Level of familiarity determined on the basis of residents' awareness of following amenities and activities planned at the park: aquarium, dolphinarium, marine museum, nature interpretation centre, rental cottages, boat and sail boat facilities, etc.
[b] 2 and 3 were combined to eliminate cells with expected frequencies of less than 5.
Source: Based on the response to the survey of the residents' conducted in Kinara Bundar, Wairy Bandh, Wairy Bhutnath, Tarkarli and Devbaugh between 4 and 18 February 2001.

traffic and noise, changes in the 'small' character of villages, increases in local taxes, disputes over land acquisition and restrictions of residents' rights regarding use of park area. The findings are consistent with the NIO study, in that the same disadvantages or impacts were identified by the study as by those who were familiar with the marine park development plan.

It may also be added that the perceived disadvantages were more than offset by perceived advantages. The opinions of better-informed residents were found to be slightly more favourable as compared with the less-informed residents. The comparison of general opinion according to residents' familiarity with the physical aspects of the project, reveal mean opinion scores (on a seven-point scale) of 5 for the better-informed group and 4.2 for the less-informed group (Table 11.8). The relationship between familiarity with development proposals and general opinion should be interpreted with caution. It is also possible that being initially in favour of the project (for reasons not necessarily related to exposure to the feasibility study) may have encouraged some residents to obtain more information. The fact that the potential advantages cited frequently by those familiar with the NIO feasibility report are those covered in detail by the same plan could explain the observed relationship between awareness levels and opinion.

Recommendations for the Proposed Marine Park

The following observations and suggestions emerge from the proposed marine park based on a field survey:

- There is a lack of sufficient familiarity of the residents with the proposed marine park. The rationale for any risks that the Malvan Municipal authority is willing to take for the marine park should be made known to all. Public awareness and interest should be created about the park by presenting the information contained in the NIO Master Plan in a more readily understandable form. Given the high literacy rates in Sindhudurg, brochures, newsletters, newspapers, pamphlets, TV, radio, etc. can be effective in summarising the salient points of the marine park plan. Popular articles and lectures in Marathi and English in schools and colleges can help in addressing the concerns already known to exist on the project.
- An examination of the park development plan reveals that it has dealt inadequately with the concerns voiced by the local residents. Comments received from residents during interviews supported the observations. Residents' rights, particularly those of the fish workers, should be clarified. Landowners in the proposed park area should be adequately compensated. Some of the possible suggestions on methods of land acquisition by the Municipality can be as follows: purchase of properties based on fair market value, long-term lease from owners, exchange of land for shares of the Municipality, etc.
- Fishing is the traditional occupation and principal means of livelihood for the fish workers who will be severely affected by the time of completion of the marine park. Although the state authorities claim that there is scope for large-scale employment, they seem to have deliberately hidden the fact that the livelihood issues of the fishermen and those dependent on fishing (fish merchants, lorry workers, ice plant workers, casual labourers, etc.) will be at stake when the marine park takes shape. Adequate and correct information on income, employment, taxes, etc. likely to be generated from the marine park needs to be provided to the residents. Above all, care should be taken to see that the financial resources raised from the project accrue to the local residents. Given the concern expressed over the possible impact of the park on local taxes and community prices, it would appear to be indispensable to ensure that residents are better informed about the financial role that the Municipality is expected to play.
- If the proposed marine park becomes a reality, its goal should be to provide for the conservation of the marine resources, education and the enjoyment of the tourists. The management of the marine environment will require cooperation between all levels of government, industry and the resident community. The following environmental safeguards are necessary to ensure sustainable development of ecotourism: the marine park should be clearly demarcated with permanent markings along the boundary limits; appropriate signboards should be erected requesting the public to help in protecting the park environment; non-renewable resources, such as sand, should not be removed from the park area and the general public should be educated regarding the

dangers from the same. The Malvan Municipality can be instrumental in determining the pace and direction of future industrial development, especially along the CRZ. Industries which are likely to release toxic effluents into the sea should not be allowed along the Malvan coast. Industrial fishing trawlers should not be allowed to discharge petroleum products (diesel, petrol, bilge oil, etc.) in the core and buffer zones of the park. Waste water treatment facilities supported by a good sewage system should be developed in Malvan, such that it restricts pollution of the sea. The Malvan Port Trust can be requested to cooperate by adhering to the rules outlined for the project. To avoid litter, particularly plastics, which have a devastating effect on coastal ecosystems, covered garbage receptacles can be kept on the beach and aboard the boats, and the tourists should be advised to use these. Regular cleaning of marine debris and beach litter by volunteers can prove particularly useful. Adoption of a policy of 're-use, reduce, recycle' can be particularly effective. In the Indian context, national park status can enhance legal protection for the park and generate marketing value and would certainly benefit the marine park (NIO, 1980).

- Responsible tourism aids in the protection of ecologically sensitive areas by minimising negative tourist impacts. Suitable rules and regulations should be framed for tourists to the marine park to prevent the collection of marine flora and fauna as souvenirs, particularly from the core zone of the park. Coral reefs along the Malvan coastline provide good opportunities for marine ecotourism activities such as scuba diving, snorkelling, wind surfing, sailing and marine wildlife viewing. Underwater inventories of fish, coral and other marine wildlife could be undertaken by the ecotourists under the guidance of trained personnel in small boats with glass bottoms and floodlights (NIO, 1980). Besides serving as an additional attraction for the tourists, this could be used effectively as an educational tool for students. An entrance fee to the marine park and a special fee for boat trips would fetch additional incomes and help in making the project self-sufficient in the long run. A concise, illustrated coastal manual could be prepared for the tourists with details on the marine park, the marine flora and fauna of the Malvan coast with descriptions and photographs. A well-equipped library could be helpful in enlisting information on the existing and endangered species of marine flora and fauna and in conducting studies on these.

Conclusion

This case study indicates the critical role played by information in public consultation. It also provides insights into the type of information that residents of small communities need for effective participation in the early stages of the public participation process. The marine park proposal has given rise to certain aspirations and aroused a number of concerns among residents, such as the impact of infrastructure development on local taxes, landowners' and residents' right to use the park zone, and so on. At an early stage of the planning process, such concerns were cited even more often than those related to the presence of tourists, most probably because their impact would be more immediate.

The overall variations between different residents' socio-demographic subgroups with respect to the concerns, aspirations and general opinion on the project were limited. Nevertheless, the associations between the various perceived impacts and the socio-demographic groups suggest residents living close to the park area and commercial fish workers are those most particularly concerned. Furthermore, it would be incorrect to assume that these groups constitute the only affected individuals from the proposed project. The owners of land in the designated areas will obviously be severely affected by the proposed park, even though they were not found to be significantly different in their perceptions and opinion from the rest of the population. These relationships need to be examined in greater detail using multivariate techniques. A detailed cost–benefit analysis on the marine park could open new avenues for discussion.

The social groups that benefit from the development process are widely different from those that are displaced by it. Moreover, while the development agents are using the local community's seas and lands as resources to be exploited economically, there is little regional or local participation in the profit derived from this. On the contrary, in most cases the local population undergoes a process of marginalisation and impoverishment. The marine park, if it evolves, will be no exception. While the development process is generally said to be a problem-solving process, displacement of people's relationships in the domains of natural resources, skills, knowledge, culture and identity is said to be a natural process. However, if development is a problem-solving process, displacement, in the name of development, could create additional difficulties for those dependent on the use of the natural resources and their traditional skills.

Meetings and discussions with the heads of the fish workers groups and other members of the *Malvan Machchimari Sangh* indicated how fishing was a respectful way of livelihood for the inhabitants of the sea coast. It is not easy for a fishing community fully dependent on the sea to give up their traditional fishing rights to adopt alternative livelihoods. Therefore, access to local resources should become a part of the development process which needs to be organised in a manner such that it bring the rules and values guiding the process as close as possible to those underpinning the livelihoods of local people. This can avoid conflict between the perceptions of the local residents and the modern development agencies, as well as make everybody equal benefactors of the process. Development cannot be achieved without the support and consent of the local population. The challenge before the developing agencies is, therefore, to initiate a genuinely participatory process involving the local people in the planning process and, more importantly, to derive wisdom from the knowledge base of the local people. In all circumstances, care should be taken to see that the profits finally accrue to the local residents.

References

Census of India (1991) *Village and Town Directory and Village and Townwise Primary Census Abstract,* Series 14, Maharashtra, Part XII A and B, District Census Handbook, Sindhudurg. Mumbai: Census Directorate.

Keogh, B. (1990) Public participation in community tourism planning. *Annals of Tourism Research* 17 (2), 449–65.

Lucas, A. (1978), Fundamental prerequisites for citizen participation in involvement and environment. In B. Sadler (ed.) *Proceedings of the Canadian Conference on Public Participation,* Vol. 1 (pp. 47–57). Edmonton: Environment Council of Canada, Edmonton.

Maharashtra Tourism Development Corporation (MTDC) (1998) *Integrated Tourism Development Plan for Sindhudurg District,* Vol. I, *Final Report*. Mumbai Tata Consultancy Services.

National Institute of Oceanography (NIO) (1980) *Proposal for the Development of a Marine Park at Malvan (Maharashtra)*. Goa: National Institute of Oceanography.

Roy Choudhury, A. (2000) Amusement parks versus people's livelihood. *Economic and Political Weekly* (9–15 Sept.), XXXV.

Sadler, B. (1979) Public participation and the planning process: Intervention and integration. *Plan Canada* 19, 8–12.

Chapter 12
Developing Sustainable Whalewatching in the Shannon Estuary

SIMON D. BERROW

Introduction

Whalewatching is defined by the International Whaling Commission (IWC) as any commercial enterprise which provides for the public to see cetaceans (whales, dolphins and porpoises) in their natural habitat (IWC, 1994). It is one of the fastest growing tourism industries in the world worth an estimated US$1 billion in 1999 and increasing at around 12% per annum (Hoyt, 2001). Whalewatching already occurs in 87 countries worldwide and is providing significant revenue for many rural coastal communities. This chapter presents a case study of an attempt to develop a model for sustainable wildlife tourism, which provides maximum benefits to rural communities while maintaining and, if possible, enhancing the conservation importance of the site.

Whalewatching in the Shannon Estuary

The concept of developing commercial whalewatching in the Shannon estuary was first discussed with the community-run West Clare Development Cooperative in 1991. During a number of surveys in the estuary between 1984 and 1991, researchers noted the presence of bottlenose dolphins (*Tursiops truncatus*). During discussions with local people, it was confirmed that dolphins (or 'porpoises' as they were known locally) were regularly observed in the Shannon estuary. In 1992, a study was commissioned by Shannon Development Ltd, a semi-state agency responsible for promoting economic growth in the Shannon region, to assess the feasibility of running commercial boat trips into the estuary to see the dolphins. This study (Berrow *et al.*, 1996) showed that bottlenose dolphins were resident in the estuary and that it was also a calving ground. Encounter rates with dolphins from boats were very high suggesting this could be one of the best locations in Europe to see dolphins. The presence of a resident group of dolphins provides the opportunity for long-term planning and investment but the study recommended that whalewatching should be marketed with other local sites of wildlife and cultural interest. It was also recommended that the development of whalewatching should also consider the conservation implications, as bottlenose dolphins are entitled to full protection under the Irish Wildlife Act of 1976.

Whalewatching in the Shannon estuary developed slowly at first, with only a small number of trips (approximately 10–20) carried out during 1993 and 1994 (Berrow & Holmes, 1999). In 1995, trip numbers increased significantly to 192 and between 1995 and 1997, about 200 trips were carried out annually from Carrigaholt and Kilrush in County Clare, catering for around 2500 people. Berrow and Holmes (1999) estimated whalewatching during this period to be worth between €108,000 and €241,000 to the local economy. During 1999 there was a 30% increase in the number of trips, and during the 2000 season visitor numbers increased by 300% to about 12,000 people (see Figure 12.1).

Tourism in Ireland was badly affected during 2001 by travel restrictions, associated with Foot and Mouth Disease in the UK. Despite this, whalewatching in the Shannon estuary continued to expand, with an estimated 17 per cent increase in trip numbers and 25 per cent increase in the number of whalewatchers.

However, there has not been a concurrent increase in trip numbers, as two new purpose-built whalewatching vessels, with three times the capacity of previous vessels, were launched during 2000 (see Figure 12.2).

A study commissioned in 1997 to examine the potential of special interest marine tourism in the West Clare peninsula identified the dolphins as the area's unique

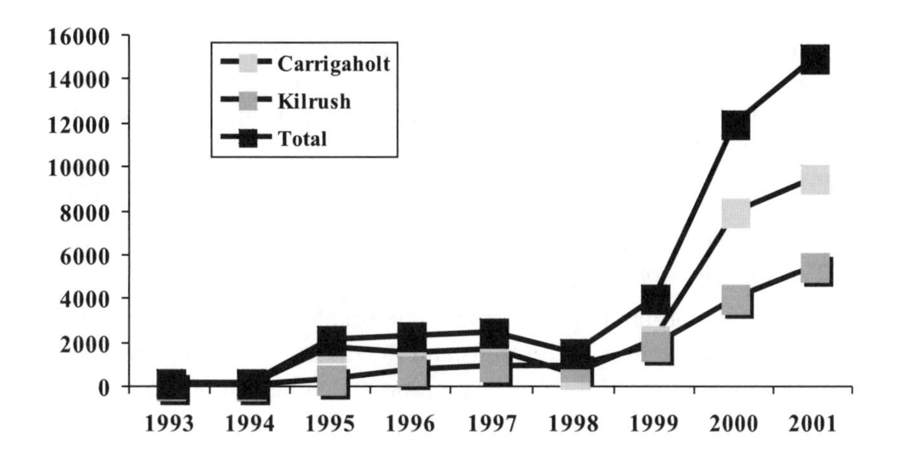

Figure 12.1 Number of whalewatchers visiting the Shannon estuary (1993–2001)

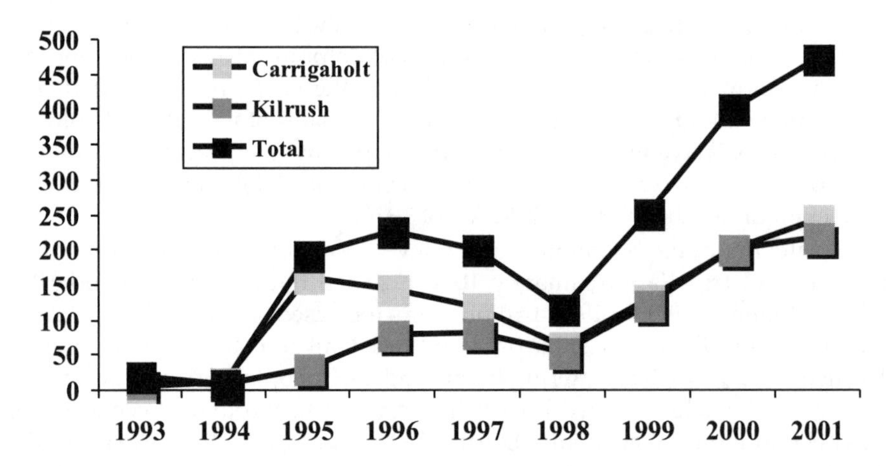

Figure 12.2 Number of whalewatching trips in the Shannon estuary (1993–2001)

tourism product (Marine Institute, 1999). It suggested that the dolphins could be the basis for an image that is special to West Clare. This study recommended that the dolphins be integrated with existing tourism products to provide a package, aimed at promoting West Clare as an activity zone, thus maximising the revenue and economic benefits to the region.

Through the Marine Ecotourism in Atlantic Area (META-) project, a brand image for marine ecotourism in West Clare was created and criteria for inclusion in ecotourism were developed (www.irrus.ie).

Framework for Sustainable Development

In order to develop genuinely sustainable whalewatching, a number of issues should be considered. These range from regulation, legislation, research and education, to stakeholder involvement and monitoring visitor satisfaction and impact (see Berrow, this volume).

In response to the rapid expansion of whalewatching in the Shannon estuary, the Shannon Dolphin and Wildlife Foundation (SDWF) was formed in March 2000, to formulate and implement a plan for the development of sustainable whale-watching. The objectives of the SDWF are:

- to maintain the dolphin population in a favourable conservation status;
- to raise public awareness of dolphins and the marine environment;
- to increase the volume and value of dolphin watching visitors; and
- to integrate dolphin watching with ecotourism activities in the region.

The SDWF is supported by a range of national, regional and local authorities and agencies including the Dúchas, the Marine Institute, Shannon Development Ltd,

Clare County Council, Kilrush Town Council, Carrigaholt Development Association and Kilrush Chamber of Commerce. The initial target is to attract and cater for 20,000 to 25,000 whalewatching visitors in 3–5 years which, using the multiplier for whalewatching in rural locations presented in Hoyt (2001), would make whalewatching in the Shannon region a £1million (€1.27 million) tourism industry and have a very significant economic impact on local coastal communities.

Implicit in the concept of any sustainable development is that the resource is not overexploited or degraded due to the promoted activity. A critical element in creating the framework for managing whalewatching in the Shannon estuary was the designation of the estuary as a Marine Protected Area (MPA) for bottlenose dolphins. Legal status for the estuary was first discussed in 1995 and a Refuge for Fauna Order drafted in 1997. However, this was never enacted. The site and species involved fulfilled the appropriate criteria for nomination as a Special Area of Conservation (SAC) under the EU Habitats Directive (1992) and thus inclusion in the Natura 2000 network, as bottlenose dolphins are listed under Annex II – species whose conservation requires the designation of SACs. The Shannon estuary was formally designated as a candidate SAC in April 2000 and transmitted to Europe as a Natura 2000 site in November 2000.

Implications of SAC Status

Under the SAC legislation in Ireland (Statutory Instrument 94 of 1997, made under the European Communities Act 1972 and in accordance with the obligations inherent in the Council Directive 92/43/EEC of 21 May 1992), the operation of commercial recreational activities, such as whalewatching, is a notifiable activity. All persons who wish to carry out this activity must obtain the written consent of the Minister for Arts, Culture, Gaeltacht and the Islands before whalewatching within the SAC. In order to obtain permission from the Minister, operators must fulfil certain requirements, namely to abide by the Code of Conduct and Conservation Plan, to provide monitoring data, and to demonstrate competence in environmental education and species identification. Both the Code of Conduct and the Conservation Plan are enforced by the *Dúchas*, the Heritage Service, which is part of the Department of Arts, Culture, Gaeltacht and the Islands.

The Code of Conduct (see Appendix), which attempts to manage the behaviour of vessels on the estuary, also applies to recreational craft using the estuary. Under this Code, operators are requested to spend only 30 minutes per dolphin group per trip, in order to minimise the impact of tour boats on dolphins and aid management of the industry. The Conservation Plan applies to the whole Lower Shannon SAC (Site Number 2165) and aims to maintain a favourable conservation status for the bottlenose dolphins in the estuary. This will be achieved through research and monitoring and assessing any negative impact on the dolphins, as well as managing whalewatching. The Conservation Plan adopts the precautionary principle, by fixing the total time whalewatching vessels are allowed in the vicinity of dolphins over the season. Following consultation with tour operators, the total time was fixed at 200 hours for the 2000 and 2001 seasons. This level will not be increased, unless the monitoring data provided by operators show that there is no negative effect on the dolphins' behaviour or their habitat.

If operators agree to fulfil these requirements, they are accredited for whale-watching in the Shannon Estuary. Under a scheme called *Saoirse na Sionna* (Freedom of the Shannon) accredited operators are awarded a dolphin flag to fly from their vessels. The concept behind *Saoirse na Sionna,* which is a type of ecolabel, lies in its promotion to whalewatchers and operators. Only accredited operators will have access to marketing and promotion from tourism agencies such as *Bord Fáilte* and Shannon Development Ltd. Whalewatchers visiting the Shannon estuary are encouraged to support only accredited operators, identified by the dolphin flag, with the assurance that these vessels are monitored and adopt good practices.

Future Challenges

At present, the whalewatching industry in the Shannon estuary is small but expanding rapidly. To develop a sustainable industry, where the resource is not degraded, we must determine the carrying capacity of dolphins to tour boats and ensure that this is not exceeded. If the target of 20,000 to 25,000 visitors is to be achieved, then assuming a typical season of 100 days and an average tour boat capacity of 32 passengers, around 625–780 trips per season or 6–8 trips per day, must be carried out. Under the Code of Conduct, vessels are requested to limit their time on dolphins to 30 minutes per group per trip. Thus, to carry 20,000 to 25,000 whalewatchers, the dolphins could be subjected to a minimum of 3–4 hours of whalewatching per day.

The ability of dolphins to tolerate this level of whalewatching is not known but if monitoring suggests that the dolphins are avoiding their preferred habitats or tour boats and the industry can be controlled, then the legal framework is available under the SAC designation. This adaptive management requires a strong scientific input. This is not only essential to ensure that whalewatching remains sustainable but can greatly enhance visitor experience through enhanced education and interpretation, provided that research is integrated with tourism.

In order to maximise the economic benefits of this tourism product to the local economy, facilities must be provided to add value to the present industry. This can be achieved through increasing the average length of stay in the area per whalewatcher. Facilities include developing onshore facilities, including land-based whalewatching as dolphins can easily be seen from headlands around the estuary, and integrating whalewatching with other tourist activities. The provision of onshore facilities, including teaching and education, should extend the season by catering for school and special interest groups. Onshore facilities can also act as an alternative activity during periods when weather prevents tour boats from going to sea or for visitors who are uncomfortable or unable to go out on tour boats. Onshore activities consolidate the industry, making it less susceptible to the weather, while increasing the average length of stay and spend per head by visitors, as well as the carrying capacity of the entire range of activities.

Acknowledgements

The Shannon Dolphin and Wildlife Foundation is supported by the *Dúchas*, Marine Institute, Shannon Development Ltd, Clare County Council, Kilrush Urban District Council, Carrigaholt Development Association and Kilrush Chamber of Commerce. Members of the Steering and Management Committees are thanked for their time, enthusiasm and commitment to this project and the concept of sustainable development.

References

Berrow, S.D. and Holmes, B. (1999) Tour boats and dolphins: Quantifying the activities of whalewatching boats in the Shannon Estuary, Ireland. *Journal of Cetacean Research and Management* 1 (2), 199–204.

Berrow, S.D., Holmes, B. and Kiely, O. (1996) Distribution and abundance of bottle-nosed dolphins *Tursiops truncatus* (Montagu) in the Shannon estuary, Ireland. *Biology and Environment: Proceedings of the Royal Irish Academy* 96B(1), 1–9.

Hoyt, E. (2001) Whalewatching 2001, Worldwide tourism numbers, expenditures and expanding socioeconomic benefits. Yarmouth Port: International Fund for Animal Welfare.

IWC (1994) Chairman's report of the forty-fifth annual meeting, Appendix 9. IWC resolution on whalewatching. *Report of the International Whaling Commission* 44, 33–4.

Marine Institute (1999) Special Interest Marine Tourism in the West Clare Peninsula. Report Commissioned by the Marine Institute, Shannon Development and Clare County Council.

Appendix: Code of Conduct for whalewatching vessels and recreational craft operating in the Shannon estuary candidate Special Area of Conservation

(1) A maximum speed restriction of 7 knots applies to an area south of a line joining the cardinal buoys Doonaha-Tail of Beal to Kilconnelly Point, as this is an important habitat for dolphins.

(2) When vessels first see dolphins they should maintain a steady course, reduce speed (< 7 kts) and monitor the dolphins. DO NOT PURSUE DOLPHINS; allow the dolphins to come to the vessel not you to them.

(3) Maintain a minimum distance between vessels of 200 m.

(4) Maximum number of three boats on a group of dolphins at any one time.

(5) Vessels on the same group of dolphins should maintain a serial course to each other if at all possible. DO NOT CORRAL BETWEEN VESSELS.

(6) Successive boats should follow the same course and come astern.

(7) Maximum time in the proximity of any one group of dolphins should be 30 minutes per vessel, per trip.

(8) New vessels into dolphin encounter zones should make VHF contact with existing vessels on Channel 8.

(9) There is to be no swimming with dolphins from commercial tour boats.

Chapter 13
Marine Ecotourism Potential in the Waters of South Devon and Cornwall

COLIN D. SPEEDIE

Introduction

On a global scale, one of the most successful tourism developments over the last 20 years has been cetacean watching, now with an annual worldwide value of over US$1 billion US (Hoyt, 2000). Other forms of marine ecotourism, which may encompass species such as sharks, seals and sea birds, have also grown considerably in recent years. This form of tourism can represent a significant opportunity for those regions fortunate enough to benefit from the presence of charismatic marine wildlife. As such, some of the wider problems endemic to coastal regions in economic decline can be addressed through the development of marine ecotourism. However, due care and attention needs to be paid to the manner in which marine wildlife is developed as a tourism product, which can be problematic if the species in question is sensitive and hence protected by law. This chapter is concerned with one such protected species, found off the coast of South Devon and Cornwall (UK) – the basking shark (*Cetorhinus maximus*).

The southern shores of Devon and Cornwall form one of the most popular holiday destinations in the United Kingdom. However, in common with many other traditional holiday destinations, there are growing pressures to diversify and enhance the visitor experience, due mainly to competition from abroad. New and attractive developments must be identified and developed to compete in a highly competitive market.

On the face of it, South Devon and Cornwall (see Figure 13.1) should be well placed to develop marine ecotourism as a new and potentially valuable opportunity. The region has a substantial existing tourism base and has a wide variety of harbours

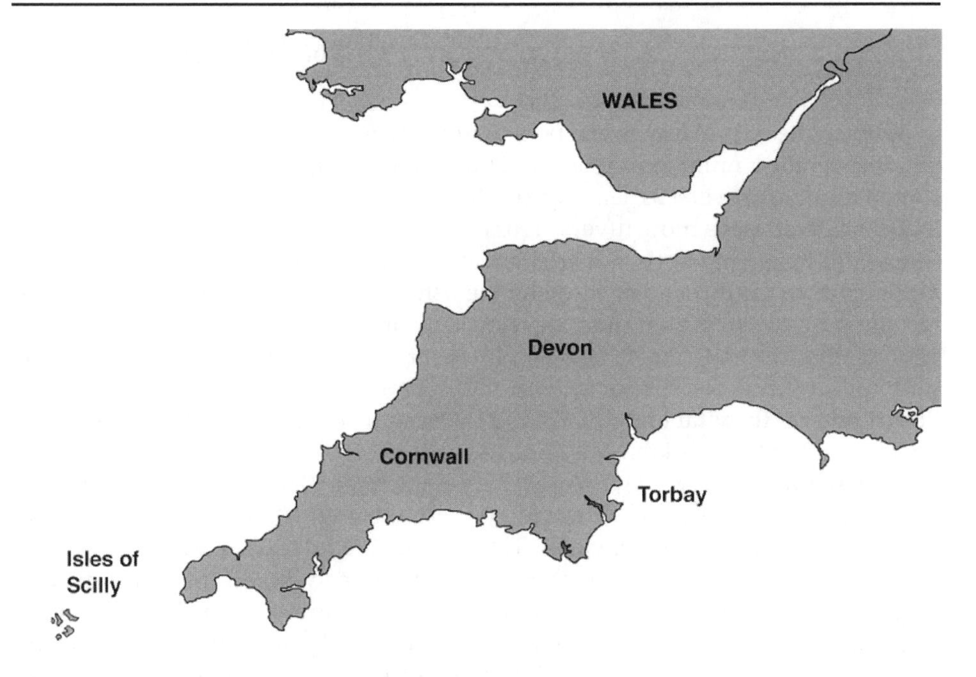

Figure 13.1 Map of South West England

now devoted more or less to leisure pursuits, together with a number of busy commercial and fishing ports, so there would appear to be a suitable infrastructure for the development of marine ecotourism ventures. Yet, so far there has been little or no development of this type.

In recent years, it has become clear that there are attractive species that inhabit the area, which if sighted often enough might support the development of suitable marine ecotourism operations. Yet experience shows that it is not enough to simply recognise and commercially exploit an available wildlife resource – issues concerning the sustainability of such enterprises must be addressed and responsibilities for the wildlife concerned must be recognised and respected. Properly organised and regulated operations should be encouraged, taking note of best practice from elsewhere, and this could allow sustainable forms of ecotourism to develop. What is to be avoided at all costs is the piecemeal development of opportunistic exploitation of a marine ecotourism resource, in which untrained and poorly-informed operators deliver a low-grade leisure experience that may, ultimately, put our often endangered marine life at risk.

In order to encourage sustainable development to be fostered, steps should be taken at this stage to ensure that training of would-be operators is available, that marine wildlife is safeguarded, and that visitors enjoy an informative and life-enhancing experience, so that a viable commercial opportunity will not be missed.

Marine Ecotourism – A New Opportunity for South Devon and Cornwall?

Many of the coastal towns and ports within the region are suffering a longstanding decline, for a number of reasons. Traditional holiday resorts like Torbay face competition from overseas, leading to a decline in visitor numbers, and are finding it difficult to attract a more diverse cross-section of the population to visit the area (Lucas, 2001). Fishing ports like Brixham and Newlyn have also suffered long-term decline due to the difficulties faced by the fishing industry. It is, therefore, widely recognised that the region must develop new attractions in order to appeal to a wider audience and to create new employment prospects. This stretch of coastline certainly has the ports, sea access, visitor numbers and infrastructure required; yet does it possess the vital ingredient – the charismatic wild creatures without which no development of this kind can hope to succeed?

For three years (1999–2001), researchers in the region have been conducting line transect surveys aboard an 11.7 m sailing vessel, observing and recording a variety of marine life during these voyages. The effort-related data gathered during these surveys has been used by a number of organisations with an interest in marine megafauna, most notably for the basking shark, a variety of whale and dolphin (cetacean) species, and rarer visitors such as the leatherback turtle (*Dermochelys coriacea*), as baseline studies of relative abundance and spatial and temporal distribution. It was recognised from the outset that such data could make a direct contribution towards evaluating the potential for marine ecotourism within the area and some preliminary findings are set out here in support of that hypothesis.

Figure 13.2 Sea Quest sailing vessel '*Forever Changes*'

The Seaquest Basking Shark Surveys: Method and Findings

A minimum of four line transect surveys of six days duration have been carried out each year since 1999, with one survey each May, June, July and August to give as broad a temporal scale as possible. Spatially, the study area extends from the Isles of Scilly (49°52.40N 06°27.00W) in the West to Torbay (50°23.70N 03°28.40W) in the East, and between these points line transects have been established, using either prominent headlands as starting / finishing points, or offshore buoys and lighthouses such as the Eddystone light off Plymouth. The lattice of transects was established to include inshore and offshore elements without favouring any potentially rewarding areas and to ensure adequate and unbiased coverage of the local waters. It was the aim of the project to complete each line transect at least twice per season.

The survey platform throughout has been the 11.7 m sailing vessel '*Forever Changes*', based in Falmouth, Cornwall. A sailing vessel makes a very practical vehicle for this type of work, being inexpensive to run, seaworthy, providing a comfortable, spacious, low environmental impact home for a crew, and capable of carrying and powering a variety of essential electronic equipment. Sailing vessels have other advantages in that they are relatively stable, slow (allowing careful observation) and 'quiet' in terms of disturbance of marine creatures, particularly those such as cetaceans that vocalise and rely on their acoustic capabilities to

Figure 13.3 Volunteer crew conducts a transect study

communicate or to locate their prey. On most occasions it has been possible to manoeuvre the vessel under sail when in contact with cetaceans or basking sharks, greatly minimising disturbance of the animals, and reducing the acknowledged risk of propeller injury if the animals close with the vessel.

On each survey a volunteer crew has been carried (minimum five) who were trained in the necessary techniques to assist with the execution of the surveys. At all times when on transect, there were two observers scanning the waters for signs of marine life to right and left of the vessel, with a height of eye of 3 m. One member of the crew was responsible for recording a wide variety of environmental data such as wind speed and direction, sea state, swell height, cloud cover and weather conditions on a half-hourly basis. Sightings were recorded on specially developed sheets to ensure that as much data as possible were gathered at all times. Position in latitude and longitude of individual sightings was established via a Global Positioning System (Raytheon R390). The ability to sight marine life is highly dependent on sea conditions and visibility. For this reason no transect was started above sea state 4 (Beaufort scale) or was continued if sea conditions deteriorated to this level, due to the greater difficulty of spotting the animals in such conditions. Similar rules were applied if visibility was poor (less than 1 mile).

Over the last three years (1999–2001) a total of 92 line transects were covered in this manner, amounting to a total of 209 hours of observation, over a distance travelled of 2088 km.

During the course of these transects, four cetacean species were recorded, the bottle-nosed dolphin (*Tursiops truncatus*), harbour porpoise (*Phocoena phocoena*), Risso's dolphin (*Grampus griseus*) and long-finned pilot whale (*Globicephala melas*).

The waters of Devon and Cornwall have few regular cetacean residents, beyond a highly itinerant group of bottle-nosed dolphins, and a poorly quantified harbour porpoise population. The common dolphin (*Delphinus delphis*) is more widespread offshore and was recorded off transect during the survey. Other species such as the white beaked dolphin (*Lagenorhynchus albirostris*) and orca (*Orcinus orca*) were also observed off transect during the study period but are believed to be irregular or seasonal visitors to the region.

No discernible pattern emerged in terms of temporal distribution, sightings being recorded in a number of different months. All cetacean sightings whilst on transect were recorded in the waters of South Cornwall.

Far more positive in numeric terms were the level of sightings of the basking shark, which was recorded on many occasions during the course of the survey.

Table 13.1 Distance per encounter 1999–2001: Cetaceans

Year	*Distance (km)*	*No. of encounters*	*No. of cetaceans*	*Km/encounter*
1999	439	2	51	220
2000	713	0	0	713
2001	936	3	54	312
Total	2088	5	105	418

Table 13.2 Distance per encounter 1999–2001: basking sharks

Year	*Distance (km)*	*No. of encounters*	*No. of Sharks*	*Km/encounter*
1999	439	12	86	37
2000	713	3	11	238
2001	936	10	37	94
Total	2088	25	134	84

Table 13.3 Temporal distribution of basking shark sightings 1999–2001

Year	*May*	*June*	*July*	*August*
1999	0	1	5	6
2000	0	3	0	0
2001	0	8	1	1
Total	0 = 0%	12 = 48%	6 = 24%	7 = 28%

Surface sightings of basking sharks are usually expected to coincide with and follow the early increase in zooplankton density, normally occurring in May or June (Sims *et al.*, 1997). Later in the summer, stratification will usually have occurred in most inshore waters of the region, where warmer water forms a surface layer over colder water, with a discontinuity known as a thermocline between the two layers. Zooplankton density will then be greater lower in the water column than at the surface, leading to a reduction in surface sightings of sharks from mid July onwards.

In order to gain a further understanding of the distribution of the basking shark population around Cornwall, analysis of the sightings made on transect reveals that with one single exception all sightings were made to the west of Lizard Point, and that the majority of sightings (48%) were made within a 9 km radius of the Runnelstone buoy (south western tip of Lands End). This was followed by sightings within a 9 km radius of Lizard Point (32%), with the remainder being sighted between these two areas (8%) or to the west of Lands End (12%). This spatial bias may be explained by the proximity of these localities to productive coastal front areas in which basking sharks are known to forage for the highest densities of their preferred prey (Sims & Quayle, 1998). Within these areas, water may remain mixed from the bottom to the surface throughout the summer, inhibiting stratification, due to a combination of strong tides, and a rapidly rising, uneven seabed. As a result, greater levels of zooplankton may be expected to migrate to the surface during all months of the summer, sustaining high levels of surface feeding sharks, which may explain the level of sightings recorded in July and August within these areas. The significance of this factor should not be ignored in terms of the critical value that these areas therefore represent for the basking shark and will have fundamental implications if any marine ecotourism operation wishes to develop within such sites.

The Runnelstone buoy is just over 16 km from Newlyn (see Figure 13.4), the nearest all-tide safe haven, and so could be reached within 1 hour by a power-driven vessel capable of 10 knots (18 km/hour), which would clearly be a

Table 13.4 Spatial distribution of basking shark sightings 1999–2001

Year	*9 km off Lizard Pt*	*Lizard – Lands End*	*9 km off Runnelstone*	*Due West*
1999	0	0	9	3
2000	2	0	1	0
2001	6	2	2	0
Total	8 = 32%	2 = 8%	12 = 48%	3 = 12%

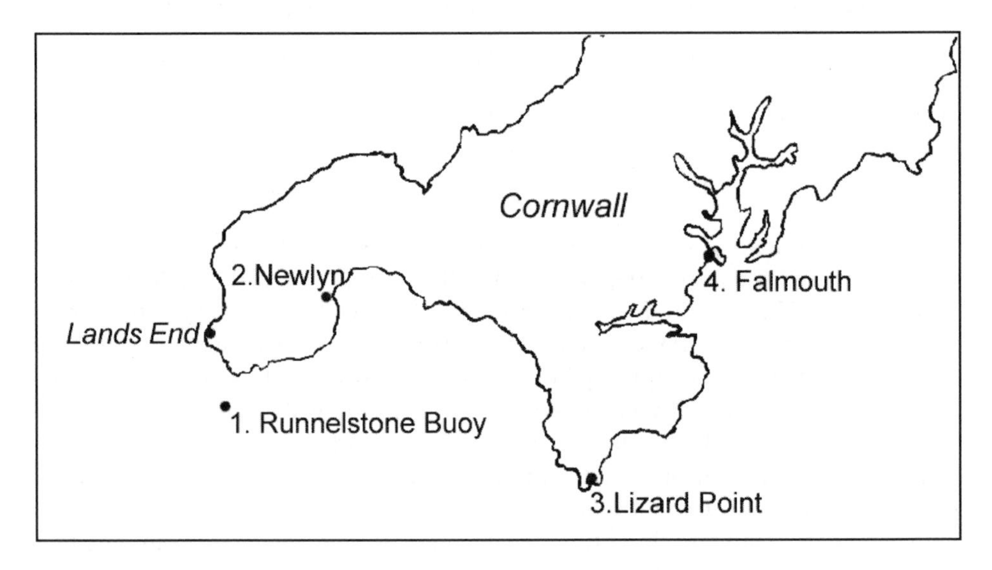

Figure 13.4 Study area for the Seaquest basking shark surveys

practical option for day charters. Lizard Point is 40 km from Falmouth and so could be reached in just over 2 hours, and thus would be equally viable for day charters.

The critical factor in the development of any commercial marine life observation activity will be the encounter rate. The figures obtained for cetaceans make a very poor case for the development of any enterprise based on whales or dolphins, with an encounter rate of one per 418 km. Assuming an average boat speed of 10 knots (18 km/hour) for a power-driven vessel this would mean that it would take 23 hours to find cetaceans which would be likely to deter even the most enthusiastic whalewatcher. Nor were the species observed of sufficient rarity to warrant special attention. However, sightings made off transect during the survey period suggest that a more dedicated boat-based study, concentrating on specific areas of the overall survey region, such as West Cornwall, might alter this picture in terms of the frequency and abundance of cetacean sightings.

The higher figures for the basking shark suggest that there may be potential for this species to form the cornerstone of marine ecotourism, at least at certain specific locations within the region, albeit for a limited season. The lack of sightings from Devon almost certainly rules out any developments in the county but West Cornwall certainly has potential. An encounter rate of one per 84 km is far more favourable and assuming the same average boat speed of 10 knots (18 km/hour) would indicate that it would take less than 5 hours to locate basking sharks anywhere within the study area. Given the spatial distribution of the recorded sightings from our survey, with all of the recorded sightings being west of Falmouth, it would be pointless and impractical to attempt to run a marine ecotourism operation based on the basking shark except within this region. By basing such an operation in either Falmouth or Newlyn the highest level of access to sharks would be assured, as with 70% of the sightings within easy reach of these two ports they must be the logical candidates for potential development.

Basking Sharks as a Marine Wildlife Attraction

There are good reasons why the basking shark might prove to be an attractive draw for visitors to observe. The second largest fish in the world, it is the biggest regular wild visitor to the British Isles, sometimes exceeding 10 m in length. There are few areas in the world in which these creatures have been recorded as regularly at the surface as in Cornwall. The other areas where basking shark sightings have been recorded include: the Isle of Man, the Firth of Clyde and the Hebrides (UK), and some areas of coastal upwellings, such as Monterey Bay, California and the New England coast (USA), and the west coast of Norway. Arguably though, Cornwall and the Isle of Man still represent the premier sites for encountering basking sharks, owing to the proximity of the sharks to the shoreline. Therefore, development of marine ecotourism in Cornwall, based on basking sharks, would certainly lend a certain uniqueness to the experience of observing the creatures closer to the shore and within a shorter time frame.

In both of the areas identified, other forms of marine life such as sea birds, cetaceans or sunfish (*Mola mola*), plus the occasional rarity such as the leatherback turtle, may be encountered. These are also Areas of Outstanding Natural Beauty (AONBs) with some of the most striking coastal scenery in Cornwall (and thus in England), which could provide a further incentive for visitors to take a voyage to see basking sharks. There may also be developments with divers, who are dedicated shark watchers, wishing to visit these areas, and, indeed, some diving operations already advertise their proximity to sites with relatively high levels of basking shark surface sightings.

The one major drawback for both of these areas is their exposed nature. Open to the Atlantic from the west, with an ever-present groundswell combined with a rising, uneven seabed and strong tides, the sea can rapidly become rough if conditions deteriorate, keeping vessels (and ecotourists) in harbour, sometimes for days on end.

Protection and conservation

Basking sharks are a protected species under Section 5 of the UK Wildlife and

Countryside Act (1981) since 1999, joining cetaceans on the list. The species is now protected under the UK Countryside and Rights of Way Act (2000) within the territorial waters of England and Wales (as are cetaceans). Therefore, any marine ecotourism operation drawing on the presence of basking sharks would have to take due regard of the law before setting up. The new Act, for the first time, specifically terms 'intentional or reckless disturbance' as an offence – a significant step, in terms of species protection. A poorly-handled commercial vessel which put the animals at risk might easily be called to account for possible infringements of the Act (despite the inherent difficulties in defining such behaviour).

In behavioural terms, basking sharks can be unpredictable in their movements and reaction to vessels in their vicinity, and so a high level of care is required when handling vessels near them. Little is known at this time of the effects of divers upon basking sharks, except that anecdotal reports from experienced film-makers suggest that, in common with some other marine species, they are disturbed by scuba gear, so snorkelling or the use of re-breathing apparatus are reported as being the preferred options when approaching the sharks.

A second factor may be that these areas of high levels of surface sightings have a significance beyond simple access to food abundance, as both breaching and courtship behaviour have been regularly observed in these areas, which may suggest that they form important gathering points for reproductive activity (Sims *et al.*, 2000a). This means that the approach of vessels or divers would have to be evaluated carefully so as to avoid disturbance of a potential breeding stock.

Photo-identification as a means of identifying individual sharks has shown that individuals may return to the area, at least on an irregular basis (Sims *et al.*, 2000b) and the current development of attaching archival satellite tags to a number of basking sharks (viewable at www.cefas.co.uk) should shed further light on this aspect of basking shark behaviour, helping to establish whether these animals display site fidelity.

If it were to be shown that the two study areas were highly significant in terms of their importance for a breeding stock of basking sharks or that substantial numbers of the sharks returned to the same site on a regular basis, then there could well be the case to argue that those sites would be worthy of further legal protection. This could involve designation under the UK National Biodiversity Action Plan (BAP) or with Special Area of Conservation (SAC) status, which would require some form of regulatory structure to be set in place for the protection of the species involved. Furthermore, the relevant authorities would have to take into account any marine ecotourism operations, and examine their sustainability in terms of their effect on the species.

A basking shark Code of Conduct (viewable at www.baskingsharks.co.uk) has been developed to offer guidance to the operators of commercial and pleasure vessels around basking sharks, but it is believed that further refinement of the code will be required as identified in the Species Action Plan (SAP) to offer more concise guidelines, especially for commercial operators of surface observation vessels or dive boat operators, to reflect increased knowledge of basking shark behaviour and reaction to disturbance, as well as the changes in the law. Some form of training and accreditation for existing or would-be operators would be essential if any form of ecotourism based around the species were not to be harmful, despite its best intentions.

In order to address these concerns, a pilot scheme will be launched in 2002 within the overall region offering training and accreditation to existing and potential masters of vessels wishing to take up some form of marine ecotourism within South West England. This scheme, entitled Watch out for Wildlife (WoW) is being backed by English Nature, WWF-UK, The Wildlife Trust, The Shark Trust, the South West Tourist Board and the South West Regional Development Agency (amongst others). It will focus on animal recognition and behaviour, the current state of the law concerning those species to be viewed, safe boat handling and the importance of developing a sound educational aspect as part of any tour. Support and guidance at this stage may well be the best thing that could happen for the wildlife, boat operators and tourists alike. Indeed, properly-trained and involved commercial operators will be a positive force for good, having a substantial vested interest in the well-being of the animals they rely upon and may ultimately be far less of a threat to the wild creatures they observe than an untrained public (Speedie, 2001).

Best practice from elsewhere will be incorporated, having particular regard of innovative and successful models such as that developed by the Shannon Dolphin and Whale Foundation in West Clare, Ireland, that are based on a sound understanding of the biological and ecological principles necessary to ensure sustainability (see Berrow in this volume). Stakeholder involvement will be encouraged at all levels, to ensure uptake, and assistance will be offered with marketing and developing the concept of marine ecotourism within the region as a further means of encouraging participation. The development of this scheme is a reflection of the fact that marine ecotourism will develop within the region come what may but experience has shown that *ad hoc* developments do not always provide the best examples of sustainable practice. This has acted as a spur for the creation of such a scheme, to provide a means of establishing a suitable platform for the development of safe and sustainable marine ecotourism within the South West England region.

It may well be that marine ecotourism operations within the region will not focus solely on whales, dolphins or basking sharks but may incorporate a variety of marine and marine-related species in combination, including sea birds and seals. Recognition of this factor should be reflected within the training and accreditation programme, which will cover a broad range of species.

Conclusion

Marine ecotourism developments dependent wholly on cetaceans do not appear to be feasible in South Devon or South Cornwall, based on the research findings outlined in this Chapter. Similarly, opportunities do not appear to exist for developments based solely on the basking shark within South Devon. However, a viable opportunity may exist in the west of Cornwall to create some seasonal developments based around the basking shark population, particularly if developed as a marine ecotourism product in combination with other marine and coastal species.

Any development involving the basking shark must pay due regard to current uncertainty concerning their reproductive behaviour, reaction to vessels or divers and vulnerability to collision with vessels. Such developments must also adhere to the new requirements enshrined within the Countryside and Rights of Way Act (2000) legislation, as well as following the existing Code of Conduct.

Further studies are planned for the next three years to continue the existing work, as well as concentrating on the two localities identified within this chapter, which will seek to elucidate many of the unanswered aspects of basking shark behaviour outlined here, particularly the further refinement of the Code of Conduct.

In ideal terms, marine ecotourism operations will become established within the South Devon and Cornwall region but will have to concentrate on a broader species base than cetaceans or basking sharks. Establishment at this time of a suitable and inclusive training and accreditation scheme could provide an impetus for the development of a valuable opportunity for the region, whilst at the same time ensuring that the wildlife necessary to sustain such enterprises is protected from disturbance. The goals of sustainability are not easily achieved in any circumstances where wildlife is involved but in the case of marine ecotourism, every effort must be made to ensure that protection of the species involved is the core value for all concerned.

Acknowledgements

The Seaquest basking shark survey was supported by grants from WWF-UK and the Wildlife Trusts, as well as contributions from the Shark Trust and the International Fund for Animal Welfare. Support was also received from the National Marine Aquarium, and the author wishes to thank all of the volunteer observers who took part in the surveys.

References

Centre for the Environment, Fisheries and Aquaculture Sciences (2002) Satellite tags help basking shark population assessment. Online document [http://www.cefas.co.uk].

European Basking Shark Identification Project (2002) Online documentation [http://www.baskingsharks.co.uk].

Hoyt, E. (2000) *Whalewatching 2000: Worldwide Tourism Numbers, Expenditures, and Expanding Socio-economic Benefits.* Crowborough: International Fund for Animal Welfare.

Lucas, P. (2001) Developing a sustainable management approach to marine ecotourism and the natural environment. Paper presented at the ATLAS 10th Anniversary Conference, Tourism, Innovation and Regional Development, Dublin October 2001. European Association for Leisure and Tourism Education ATLAS.

Sims, D.W., Fox, A.M. and Merrett, D.A. (1997) Basking shark occurrence off South-West England in relation to zooplankton abundance. *Journal of Fish Biology* 51, 436–40.

Sims, D.W. and Quayle, V.A. (1998) Selective foraging behaviour of basking sharks on zooplankton in a small-scale front. *Nature* 393, 460–4.

Sims, D.W., Southall, E.J., Quayle, V.A. and Fox, A.M. (2000a) Annual social behaviour of basking sharks associated with coastal front areas. *Proceedings of the Royal Society* B 267, 1897–904.

Sims, D.W., Speedie, C.D. and Fox, A.M. (2000b) Movements and growth of a female basking shark re-sighted after a three year period. *Journal of the Marine Biological Association of the United Kingdom*, 80 (6), 1141–2.

Chapter 14
Scuba diving: An Alternative Form of Coastal Tourism for Greece?

CHRISTOS P. PETREAS

Introduction

Scuba diving, as a form of tourism activity, has tended to attract the 'higher spending tourist'. Yet Greece, a country of extended coastal areas and many islands, has not been known for or promoted this type of tourism. Only in the last 10 years, following public pressure from interested organisations and individuals, has scuba diving changed from being a 'prohibited activity' to becoming a popular sport that is offered to tourists.

Greece is still a long way removed from promoting itself as a 'scuba-diving paradise', even though some areas have been de-regulated and tourist scuba diving is allowed. The activity is still raising much controversy between the various diving professionals, the tourism associations, and the government (for example the archaeological authorities).

This chapter examines the marine ecotourism environment of Greece, assesses the present framework and practices of scuba diving and presents the results of a recent survey among dive operators in Greece.

The marine/coastal environment in Greece

Marine ecotourism has been defined as 'nature-based tourism that interprets marine wildlife, environment and local marine/coastal culture and heritage, providing a quality experience for tourists, while minimising the impact of tourism on the marine environment' (Wilson *et al.*, 2001). The development of coastal/marine tourism should, therefore, include activities which are not only economically beneficial to the coastal area but also take into consideration the importance of the marine life and marine environment more generally, focusing on activities that support the sustainability of such environments.

Greece is blessed with an extensive and abundant marine / coastal environment (Katsadorakis, 1999). Greece, a country in the southeastern Mediterranean with a total shore length of 16,575 km (out of the approximately 46,000 km of Mediterranean shoreline), has over 9000 large and small, inhabited and uninhabited, islands, among which are the well-known tourist islands of Crete, Rhodes and Corfu. The deepest point in the Mediterranean (at 5093 m) is in the southern Ionian Sea to the west of Antikythira island.

Approximately 77% of the total fish species of the Mediterranean (447 out of a total of 579) have been recorded in the Greek seas, among these are about 75% of the various shellfish species. Fifteen of the sixteen types of mammals appearing in the Mediterranean have also been reported.

The Greek seas are relatively poor for fishing. The eastern Mediterranean produces only 15% of the fishing product of the Mediterranean as a whole, while the western area produces 38% and the central area 47%. However, the lower abundance of fish results in the clear blue-green waters which attract tourists (Katsadorakis, 1999). Musa (in this volume) refers to a number of studies which consider underwater visibility to be the second most important factor in tourists' satisfaction with diving experiences, the most important being the presence of marine life.

The Greek coastal zone (including the islands) accounts for about 60% of the country's population, 40% of agricultural production and 70% of industrial activity. However, only one of the 37 swimming coastal areas was reported to have slightly higher than the standard microbe factor measure (Katsadorakis, 1999). Greek seawaters are monitored continuously for all types of pollution and safety for swimming by measures put in effect by the Ministry of Environment.

Greece's first marine park was established in 1992 for the protection of the Mediterranean seal, which is said to be one of the six most endangered species of mammals in the world (it is estimated that there are only 400 to 600 individuals left alive, of which 200 to 250 live in the park area). The Marine Park, now designated a National Marine Park, includes 19 small islands (most of them uninhabited, the total population of the area covered being about 3000 people), which are spread across an area some 1587 km^2 in size. The Park has one central access point: the small island of Alonissos in the Northern Sporades, off the coast of Evia island. Although initially the major economic activity of this area was fishing, with a little agriculture, in the recent years tourism has taken over (Katsadorakis, 1999).

In 2000, a new National Marine Park was established off Zakynthos island, the habitat of the Loggerhead (*caretta caretta*) (Pantis *et al.*, 2001). The Loggerhead turtle is one of the oldest forms of life on the planet and 2700 of the total estimated 4500 population come to lay their eggs on about 100 km of the Greek southern coast, the most important area being in the gulf of Laganas and on a 300 m coast of Sekania, in Zakynthos (Katsadorakis, 1999). The new National Marine Park was set up with funding assistance form the EU Life Programme and is to be managed by its own organisation. It proposes to collaborate with both the local authorities and the international tour operators, who bring over 400,000 tourists to Zakynthos annually.

The sport of scuba diving

Among the specialist activities of coastal tourism, the sport of scuba diving has been one of the most popular, with an increasing trend in recent years. According to the Professional Association of Dive Instructors (PADI), the world's largest scuba-diving certification organisation, the number of certified divers worldwide increased from around 2 million in 1987 to 8 million in 1997. These figures do not count the doubtless many more who take part in 'try-dive' programmes and those who practice snorkelling (use of mask but without breathing apparatus).

Shortly after the Second World War, the invention of the autonomous breathing device provided humans with the opportunity to make a very old dream come true: the ability to dive safely and get to know first-hand a world until then unexplored and mysterious. After being perfected so as to meet the highest standards of safety, the autonomous breathing device became part of the equipment of amateur divers, whose number started increasing rapidly. Nowadays there are several million of them in the USA and in the developed countries of Europe.

The constantly rising numbers of amateur divers and the increasing demand for new places to dive has resulted in the development of a new, special form of tourism, which has become very popular all around the world in a relatively short time. In economic terms, this has given rise to manufacture of the required equipment, training for the sport and the putting together of specific diving tour packages. The Caribbean region provides 57% of scuba-diving tours worldwide, which are expected to earn $1.2 billion annually by the year 2005 (Townsend, 2001).

Diving tourism is considered to be one of the most dynamic and profitable forms of tourism, since it is a hobby of millions of scuba divers of a medium or high income who travel all around the world, combining scuba diving with their holidays. In the meantime, thousands of tourists learn the basics of how to dive in the destination where they are having their vacation.

There are many countries that capitalise on the magical appeal that the sea world has on everyday people and have directed efforts, in many cases successfully, to develop diving tourism and to realise the financial benefits of this tourist activity. The main international-level scuba-diving destinations are the Caribbean, the Red Sea, Thailand and Australia's Great Barrier Reef. Other destinations are seeing the potential and are developing scuba-diving tourism, among them Scotland, Malta and South Africa. Countries less favoured by nature in this field, such as Switzerland and Germany, have grown into considerable centres of diving tourism, thanks partly to their fully-equipped diving centres but mostly due to their will to promote and advertise this sport. From the point of view of natural beauty, as well as geographical and climatic characteristics, these countries are much less favoured as far as diving is concerned than coastal countries (such as Greece). Yet, thousands of divers dive every year in the cold, muddy and sometimes unfriendly lakes and seas of northern European countries. This is because there are few alternatives that can satisfy their desire to explore the underwater world (Papasimakopoulos, 2000).

Most of the countries in the Mediterranean basin have taken advantage of the rising interest in scuba diving and have developed the necessary facilities to satisfy the tourist customers' needs. In addition, Mexico has expanded into offering diving even in the relatively shallow areas of the Yucatan peninsula cave system.

Coastal Tourism in Greece

The development of tourism in Greece

Modern Greek tourism began in the 1950s and was developed primarily as coastal / resort and cultural tourism, in that the government interventions of that period concentrated on the construction of infrastructure in the coastal areas, near to known sites of historical / archaeological and cultural events.

Throughout the early 1970s the growth in Greek tourism was spectacular, with the exception of the 'crisis year' of 1974, which saw the fall of the junta military government in Greece and the Turkish invasion of Cyprus. In 1975, Greece made up everything it had lost the previous year, and in 1976 Greek tourism expanded by some 34% to take international arrivals to 4.24 million. By the end of the 1970s, however, growth had slowed down. Since that time it has been unspectacular and patchy, only breaking the 10 million arrivals threshold in 1994 and 1995. The country experienced slow growth after 1996 and 1997, which were depressed, and reached approximately 11 million arrivals in 2001. Income from tourism regularly amounts to 40% of invisible earnings and covers 35% of the international trade deficit (Hellenic National Statistical Service, 2001).

In the first decades of Greek tourism, the emphasis was on developing infrastructure and facilities in the coastal areas and in some of the (now) more popular islands. Some interventions supporting tourism by way of facilitating investment, mainly in accommodation, contributed to this process of development. In the latter part of the 1960s the main casinos of Athens (Mount Parnes), Rhodes and Corfu were put into operation.

In the mid-1970s the first efforts to develop specialist facilities were undertaken with the ski resorts of Central Greece (Mount Parnassos) and in Macedonia. In the same period the first rural tourism interventions were the reconstruction of traditional residences in the Mani and other regions.

In more recent times government intervention has also promoted the development of health tourism facilitates in some of the more popular spas and a number of mountain refuges (in collaboration with the Alpine Federation of Greece). The coastal areas have also been re-emphasised with a number of port facilities and yachting marinas.

However, in spite of all these developments and the increasing tourism flows, from the beginning of the 1990s Greek tourism began to suffer from a multitude of

Table 14.1 Distribution of tourism arrivals in Greece by country of origin as a percentage of total arrivals for selected countries

Country of origin	*1995*	*1996*	*1997*	*1998*	*1999*
United Kingdom	20.8	17.3	17.0	18.7	20.0
Germany	21.2	19.5	19.8	19.6	20.1
Italy	6.0	5.0	5.3	6.0	6.1
France	5.2	4.7	4.2	4.5	4.5
Scandinavian countries (Sweden, Norway, Denmark, Finland)	9.6	10.5	11.2	10.4	10.4
Low countries (The Netherlands, Belgium, Luxembourg)	7.0	6.8	6.9	7.5	7.8
USA	2.2	2.3	2.4	2.0	1.9

difficulties (Tsekouras *et al.*, 1991). These include uncoordinated efforts on the part of its many organisations (governmental and professional), inconsistencies in government tourism policy, the lack of specific tourism infrastructure and facilities and insufficient levels of investment in tourism.

It is logical that with a long period of sunny days and mild temperatures, tourism in Greece has been developed primarily on the coastal areas and among the islands. Greece, a popular destination with tour operators, now receives around 5.5 million package tourists annually (Research Institute for Tourism, 2001: estimated 2001 figures). Being the main tourism activity, coastal tourism (often described as '3S' tourism, reflecting its basis on sun, sea and sand) is the major breadwinner of the Greek tourism industry. Until recently, 90% of the tourist facilities were located in coastal zones and the islands. The Greek tourism industry relies heavily on organised, operator-directed mass tourism, primarily from the United Kingdom and Germany, which together account for about 40% of the total. Greece is also popular with Scandinavians and in the recent years increasingly with tourists from the former Eastern Bloc countries.

In Table 14.1, visitor arrival statistics for the main origin countries are presented. They are important in that they have a bearing on scuba-diving activity (see analysis of characteristics later in this chapter).

Interest in 'alternative' forms of tourism

The decade of the 1990s saw a growing interest in further developing the tourist product, mainly because while mass tourism was increasing in volume it was contributing a smaller revenue per tourist capita. This also created pressures for cost reductions on the part of tour operators. Furthermore, a number of traditional coastal tourism destinations experienced a decline in arrivals and became concerned with the need for less dependence on the seasonality of the 'sun–sea–sand' syndrome. They were also affected by increased competition from new destinations.

The main emphasis of recent national-level tourism development policies has therefore been in extending the tourism season and improving the quality of the tourism product. Related to these main goals have been efforts to attract higher income-generating tourists; increase the average accommodation occupancy; develop new, so-called 'alternative' forms of tourism; and achieve a better distribution of the tourism activity, both geographically and across the tourism season.

Tourists' interest in 'nature-related' tourist activities and the provision of EU funds for the support of income-generating activities in the economically poorer areas – primarily rural areas – offered opportunities for tourist products in the mountain regions, in ecologically-related activities and in agro-tourism. Sustainability issues have also been emphasised in the recent tourism developments. Areas with practically no tourism activity are becoming tourist attractions based on their natural resources.

The government, therefore, instituted a number of policies to treat the symptoms: a new incentives law to support accommodation installations with specialised facilities such as skiing, conference centres, private marinas, health spas; rural and agro-tourism are also supported. European initiatives such as LEADER are emphasising tourism-related developments and there has been a major effort to include interventions for tourism activities in inland and mountainous areas.

Coastal tourism activities, *per se*, have not been in the forefront of government intervention, except for an emphasis on improving the tourism product offering and adding ancillary facilities (such as conference centres, golf courses, swimming pools, etc.) to existing installations or businesses (Government Law No.2601/98).

Coastal areas with a heavy concentration of accommodation facilities were designated as 'full' and new investments were excluded from the new incentives law. However, it became evident that these areas would continue to attract large numbers of tourists and a different approach has been developed recently. The emphasis now is in enhancing the tourism product with additional services, new facilities to help extend the traditional tourist season and the promotion of ancillary attractions.

The emergence of scuba diving

Among the various tourism activities, 'sports or athletic tourism' includes either the participation in, or the watching of, sports events or activities. In recent years, with the increased interest in sports and adventure tourism, a number of activities have been described as 'alternative forms of tourism' in contrast to the coastal '3S' tourism. Within the definition of sport tourism activities are such sea activities as water skiing, snorkelling and pleasure scuba diving.

These activities, while coexisting within the offering of the coastal/resort tourism, have not normally been promoted separately and have traditionally been included either in the 'facilities' descriptions of various resorts or as optional activities in a 'sea resort package'. In this context, they have not been the objects of any particular examination of the interest of tourists by any of the official organisations.

Indeed, the activity of underwater diving – scuba diving – has not been considered either separately or individually by any official organisation. It has not even been mentioned as part of the 'new forms' of tourism, despite the fact that the geographic nature of the country with an extensive coastline and many islands would justify its categorisation as such.

There are three main reasons for this:

- Most important is the fact that no free diving has been permitted because of the need to protect underwater archaeological finds and the fear of either destruction of the sites or unauthorised removal of the items.
- Greece had not been developed as a 'diving' destination and had not created an image for this either locally or internationally.
- The framework governing the operation of scuba-diving activities (both in terms of training and in terms of offering pleasure dives) has been unclear, difficult to satisfy and complicated to apply.

Thus, little scuba activity existed in Greece up to 10 years ago and there were few Greek divers or diving schools, although some foreign tourists and some locals were practicing scuba diving, often unauthorised and in locations that were officially restricted (Anon: 1993).

The Framework for Scuba diving in Greece

The official and legal framework

Until the mid-1990s, scuba-diving activities in Greece were covered by the general regulations included in the Port Operation Code. The inefficiency of these regulations had been recognised as far back as 1985 but efforts to update and create an appropriate legal structure were fragmented. Eventually, in 1993, the then Minister of Culture signed a decree making pleasure diving possible in 10 locations around the country and giving the various diving clubs and Greek divers the satisfaction of being able to practice their sport legally (Argyrakopoulos, 2000).

In 1994–95 a series of exchanges and collaborations took place among the relevant competent authorities and organisations, primarily the following organisations:

- the Ministry of Merchant Marine;
- the Secretariat General of Sports;
- the Hellenic National Tourism Organisation;
- the Ephorate for Underwater Antiquities of the Ministry of Culture, and
- the Hellenic Federation of Associations of Underwater Activities, Sport Fishing and Swimming Techniques.

The result of these meetings was the creation of a new legal structure for underwater activities, which was put into effect in January 1997 and was added to by special regulations in April 1999 (Argyrakopoulos, 2000). The regulatory framework is thus relatively very new in Greece.

The new framework distinguishes between 'training centres', which are for learning underwater swimming with an independent breathing apparatus (effectively pleasure scuba diving), and 'diving centres', which operate a craft for the purpose of organised pleasure dives by certified divers. The framework also defined the requirements for a 'diver training instructor' and designated the tourist season for pleasure dives as 1 April to 31 October each year.

The individual coast guard authorities are involved in the process of issuing licenses to the various training and diving centres, the licenses being valid only for up to two years (expiring on 31 December of the year following their issue). The initial issuing of a license is dependent on the centre having the relevant and required facilities, a craft and necessary personnel duly trained and certified. In October 2001 the then Minister of Merchant Marine, in a public presentation, presented the basic points of the new governmental action plan for the development of touristic diving in Greece (Ververis, 2001). Among the ten points which will form the core of the new legislation of interest were:

- the delineation of areas and creation of new underwater marine parks,
- the provision of grants for the scuba-diving training schools and diving centers and
- The establishment of a joint programme with the Ephorate for Underwater Antiquities in order to 'de-regulate' additional areas for pleasure diving.

The official entities that are authorised to issue Greek scuba-diving certifications (similar to international one-, two- and three-star divers) are:

- the Ministry of Merchant Marine,
- the Secretariat General of Sports,
- The Hellenic Federation of Associations of Underwater Activities, Sport Fishing and Swimming Techniques,
- the Unit for Underwater Demolitions of the Hellenic Navy,
- the Unit for Underwater Missions of the Hellenic Coast Guard, and
- certified (by the Ministry of Merchant Marine) Scuba-diving Training Centres.

Of course, the certified training centres also offer international certification (see the section on characteristics of scuba-diving services, below).

The currently applicable regulations for pleasure diving (which are indicative only) permit diving with breathing apparatus from sunrise to sunset. Diving is permitted only in areas that have been so deregulated by the competent authorities (basically the Ephorate for Underwater Antiquities of the Ministry of Culture). The regulations also restrict diving in areas where antiquities are located or found by divers (who are required to report to the local coast guard authorities any such finds). The authorised pleasure diving depths are 18, 25 and 30 m for the holders of the respective one-, two- and three-star diver certifications. There is a series of additional regulations that must be applied for safety and other reasons (Argyrakopoulos, 2001).

Practical aspects

In terms of activity, the sport of scuba diving is included under the coordinating responsibility of the Secretariat General for Sport, which is part of the Ministry of Culture. Meanwhile the athletic responsibility rests with the appropriate sports federation (there are about 40 federations having responsibility for different sports), in this particular case the Hellenic Federation of Associations of Underwater Activities, Sport Fishing and Swimming Techniques. The Federation has approximately 120 member associations all over Greece. A significant point to note is that the Federation's activities are so broad that scuba diving is really only a small part of its activities. Member associations are authorised to practice scuba diving for their members (including training and organising dives). However, it is not possible to assess the extent of such activities in the absence of a detailed survey of these associations. Scuba diving may be one of many interests or activities undertaken, and association members may be involved in more than one of the activities of their association (i.e. scuba diving and others). More specific to scuba diving is the Association of Trainers for Professional and Amateur Diving, which has approximately 150 members. They are either owners or staff of the various certified diver training centres and pleasure diving centres. Their main interest is as a professional association. There are, therefore, no scuba diving 'clubs' in Greece, as there are in other countries. The centres organise various scuba-diving excursions for their clients and other parties. The latter are primarily addressed to local divers, mostly ex-students of each centre.

In early 2002, a number of centres decided to form an association of centres, which was intended deal more specifically with the problems of the organisation of training for and organisation of pleasure diving and to promote the deregulation of more sites for diving.

Diver training is in two main categories: pleasure diving and professional diving. Pleasure diving, being the more popular, includes the basic certifications of one-, two- and three-star diver training and specialist training in wreck diving, underwater photography, mixed air (for deeper) diving and cave diving. Professional diving training is for industrial, marine, military and special missions.

Most tourist pleasure diving is done in areas where coastal tourism is already active and for most visitors it is offered in one of three types:

- Part of a tourist package that is prepaid by the tourist. This type of product covers usually a full day of activity, with some training, a local visit and a short familiarisation dive (maximum half hour). It does not require certification.
- Purchased locally or at the resort (some centres are associated with a particular resort). This type may include full training with certification and standard (1 hour) and repeat dives. Equipment is included in the overall package price.
- Purchased locally or at the resort by already certified divers. This type includes sometimes a two-day diving excursion by boat. It may or may not include the rental of equipment.

Problems

The main problems in Greece are identified with the operation of the training and diving centres and with the provision of scuba-diving services. They can be grouped in two categories: problems arising out of the legal framework that regulates the provision of these services; and problems arising out of the lack of specific interest by the tourism authorities in the promotion of the scuba-diving activity as an integral part of the Greek tourism product. It is worth noting some of the specific problems (based on author interviews, 2001–02):

- Collaboration between the Ministry, the Federation and the Trainers Association, is still at an embryonic stage.
- The work of the 'diving trainer' is not recognised as such and at present trainers are categorised professionally as 'instructors of sea sports'.
- The legal framework hinders the financial viability of the 'pleasure diving enterprise', thus pushing the operators either to do it as a side business, to emphasise equipment sales or to do it without proper compliance with regulations.
- It is probable that a number of diving services are offered by unlicensed foreign trainers who come to Greece seasonally and cater for tour operator clients.
- There is no official insurance coverage provision for the diving centres, each operator being left to make individual arrangements.
- There is little if any documentation or guidebooks for Greek underwater sites.
- Greek university courses on the subject of oceanography do not include diving training for their students.
- There is no national action for promotion of the sport.

The recent initiative by the authorities (see earlier) will begin to address some of these problems but more collaborative work is needed in order to bring this sport to the level it requires.

Present Scuba diving Activity in Greece

Underwater activities, primarily individual scuba diving, underwater photography and snorkelling, have become more popular in the coastal locations and in the islands in the last seven or eight years, particularly with the official 'deregulation' of a number of areas allowing scuba diving. While international scuba diving has seen substantial development and is being practised even as a weekend sport, in Greece, with the exception of the main cities most scuba diving is offered at or near tourism resorts.

Scuba diving offered in Greece can be grouped in three categories:

- the training and leisure diving activity offered by training centres in the Athens and Thessaloniki area (and a couple of other larger towns), which is directed to residents as a form of 'adventure' or sports activity;
- the training and leisure diving activity offered by various centres, primarily

located in sea resort areas and large hotel complexes and available for the tourists there (either hotel residents or visitors in the area); and

- the leisure diving activity offered in a number of location for certified divers, either Greek residents or tourists (foreign or Greek), in a number of locations, by diving centres or by combined training/diving establishments.

Geographical distribution of scuba diving activity

Based on data from the Ministry of Merchant Marine (and Argyrakopoulos, 2000), the responsible authority for licensing underwater diving schools and underwater diving centres, there were 88 licensees at the end of 1999 in Greece, out of which it is estimated that about five ceased operations in 2000. They are distributed in a number of areas in the country – directly related to the authorised diving locations – and were the following at the end of 1999 (see Table 14.2).

Table 14.2 Distribution of diving schools and centres in Greece

Location	*Training Centre*	*Diving Centre*	*Training and Diving Centre*	*Total*
Patras	1			1
Ag. Nikolaos – Crete	3	4	2	9
Volos	1			1
Zakynthos		9		9
Heraklion – Crete	3	1	1	5
Hydra Island		1		1
Thessaloniki	4	1		5
Santorini		1	1	2
Kavala	2			2
Kalymnos		2	1	3
Kerkyra (Corfu)	5	7		12
Kefallonia	2		1	3
Lefkada	1	1		2
Mykonos	1	1		2
Paros		2		2
Athens - Piraeus	8	1	1	10
Porto Cheli – Peloponese		1	1	2
Rethymno – Crete	4	2		6
Rhodes	3	1		4
Sitia – Crete	1			1
Skiathos		1		1
Chalkida	2			2
Chania – Crete	2	1		3
Totals	45	35	8	88

Figure 14.1 Geographical distribution of training and diving centres in Greece

We see from the table that 51% of centres are training centres, 40% are diving centres and 9% are combined centres. The centres are mainly located in the islands on Crete (27.3%) and on the other islands (43.2%), while only 27.3% are on the mainland.

The centres are located in 23 different locations in Greece, in five locations on the island of Crete (the largest concentration) plus 11 other islands, with another seven mainland locations. Their locations are identified in Figure 14.1.

We should note that at present three important regulations affect diving activity:

- Diving is permitted only in the designated areas in different locations in Greece, which have been 'deregulated' (or freed from the archaeological restrictions) by the Ephorate of Underwater Archaeology of the Ministry of Culture.
- Night (between the hours of sunset and sunrise) diving is not permitted.
- Spear fishing or other fishing while wearing scuba equipment is forbidden.

There is one positive (but often loosely interpreted) regulation which helps the tourist product offering: it is permitted for persons not certified as divers to practice a so-called 'discover scuba' dives, at depths not exceeding 5 m, for a period of up to

10 minutes. These are for one time only and for the express purpose of acquainting participants with underwater diving.

As in other parts of the world, the regulatory framework for scuba training requires a medical examination and statement to be undertaken and for would-be divers to have both theoretical and practical training before certification is awarded. The Greek authorities recognise international diving certifications and offer a 'Greek' certification, similar to the others, of one, two and three stars.

Characteristics of scuba diving services

In order to assess the present situation of the 'scuba-diving tourism product', the author carried out a survey among the authorised scuba training centres and the diving centres. A questionnaire was distributed and a follow-up telephone call was made in order to obtain a 34% response rate. This response rate is actually estimated to be closer to 40% since it is understood that some of the listed centres (based on the Ministry of Merchant Marine 1999 list) were no longer in operation at the time of the survey.

The survey was considered preliminary in that it became apparent following the responses and the telephone discussions with some of the owner/operators that a more in-depth examination of the scuba-diving activity as a tourist product should be undertaken in the future (Petreas, 2001).

The questionnaire included four sections of questions: some basic characteristics of the centre, details of the type of services offered, some general information on the clients, and an indication of the problems and prospects for the future.

Of the responses received 27% originated from Crete, 43% from the other islands and 30% from the mainland.

The date of original establishment ranged from 1982 (oldest) to 1999 (most recent) with an average length of operation of seven years. Around 33% of the centres operate year-round, while the rest operate on a seasonal basis, on average for seven months a year. A variety of services are offered, as shown in Table 14.3. In terms of training, different Greek and international certifications are offered. The

Table 14.3 Services offered by scuba-diving centres

Type of service	*% of centers offering*
Scuba training	86
Accompanied (guided) dive	100
Unaccompanied dive	40
Equipment rental	60
Tank filling	80
Transport by car	83
Transport by boat	96
Other service (discovery dives, rebreather dives)	6

Table 14. 4 Scuba training certifications offered

Type of certification	*% of centers offering*
Greek certification	16.7
PADI	70.0
CMAS	56.7
Other (SSI, NAUI, INTD)	16.7

details are shown in Table 14.4, which shows that over half of the centres offer more than one choice of certification.

The types of clients of the Centres are both Greek and foreign. On average they indicate 25% Greeks and 75% foreign clients, indicating that they primarily address foreign tourists rather than Greek tourists or the residents of their area. This is backed up by the finding that the Greek clientele represents up to 5% of all clients in 36.7% of all the centres, while the foreign clientele is at 90% or over in 50% of the centres surveyed.

The length of the training offered ranged from as little as 16 hours to as much as 45 hours, with an average of 31 hours. The centres surveyed charged from €205 to €470 for the training course, with an average of €314.

The usual diving activity is done in dive excursions. Most centres offered two-dive excursions with an average cost of €65. Alternatively they stated that they also offered individual dives at an average cost of €40.

The distribution of the foreign clients' origins (they were asked to state up to three origins) is shown in Table 14.5, where the percentage of overall tourism arrivals from the same country is shown (see also Table 14.1).

In terms of the problems holding back the development of the product offering, 96.7% felt that there was inadequate promotion, while 100% felt that the authorities should expand the unrestricted areas.

There was a general consensus that there is potential for attracting more clients. Ninety per cent of the centres surveyed indicated they felt there was potential for more clients in general, 86.7% felt there was potential for more Greek clients, while slightly more, 90%, felt there was potential for more foreign clients. It is interesting to note that 10% responded that they did not feel their clientele could expand. However, no explanations were given on the reasoning behind this position.

As a final comment, it should be noted that in some more popular resorts, information was received about 'unauthorised' diving activities offered to tourists within the services of a resort hotel complex. The researchers were not, however, able to obtain any data on these activities. One reason for this is that these 'centres' are not listed nationally. Another is that some may have obtained a temporary license for operation from local coast guard authorities. This may either be unconfirmed or not renewed, in which case they would probably never find their way into the official list of the Ministry of Merchant Marine. Third, such diving activities are often carried out (quite legally, since they are not required to obtain a separate

Table 14.5 Foreign clients of scuba diving by country of origin

Country of origin	*Percentage of centres stating this nationality among their three most important foreign clients*	*Proportion of Greek tourist arrivals1999 (Table 14.1)*
United Kingdom	46.7	20.0
Germany	76.7	20.1
Italy	6.7	6.1
France	13.3	4.5
Scandinavian countries (Sweden, Norway, Denmark, Finland)	30.0	10.4
Low countries (The Netherlands, Belgium, Luxembourg)	–	7.8
USA (in Mykonos and Santorini)	6.7	1.9
Austria	6.7	–

license) within the activities of athletic associations or clubs related to sea, sailing, swimming and fishing activities.

It is interesting to note that an overview of diving destinations listed in British diving magazines (Anon., 1998) relates primarily to the diving centres located in the popular destinations of the particular nationality (Corfu, Crete, Zakynthos). Of course the listings are usually the result of promotion of the individual operators and cannot therefore be taken as general indications of what the foreign visitors know about the diving areas of Greece.

Conclusions

Assessment of the present status of scuba diving in Greece

The geographical characteristics of Greece are a positive asset to the enhancement of the coastal tourism product with scuba-diving activities. It is clear from the survey results that the foreign tourists do engage in the scuba-diving activity and the product has potential for expansion. Indeed, the promotion of this activity is still embryonic and its absence hinders the use of scuba diving as a core attraction. Participation in international scuba diving exhibitions is, at best, incidental and scarce.

The regulatory framework for the activity clearly does not sufficiently cover all of the various aspects of scuba diving but according to official sources it is in the process of being revised. Few areas have been designated as 'open diving sites', primarily due to the restrictions for the protection of Greece's historic heritage and archaeological finds.

Diving services for tourists seem to be adequate given the present level of activity, and there is cognisance on the part of the operators of the services of the needs of the tourists. Also, the factor of 'crowding' (described by Musa and Townsend, both in this volume) is far from a problem in the Greek context.

Environmental considerations

As noted earlier, until recently only one formally-designated protected area, for the Mediterranean seal, existed in Greece as a 'marine park', although a number of coastal areas have been designated as 'restricted to various activities' in order to protect different species of marine life. Among the best known of the latter category is the Zakynthos coast, the breeding ground of the Loggerhead turtle. Planning issues for scuba diving have not been part of formal tourism policy issues so far, since scuba diving has not been actively supported. However, planning for marine ecotourism, as indicated earlier, has recently been started in Greece, with the designation of the Zakynthos Marine Park.

Under a new initiative by the Ministry of Merchant Marine, some of the planning issues will be addressed. However, there is still much work to be undertaken in order to develop scuba diving as a marine ecotourism activity, particularly in respect of local participation, the adoption of a collaborative approach and monitoring (Wilson *et al.*, 2001).

Divers generally express great concern for the environment, particularly the cleanliness and pollution of the shore area and underwater (Musa, this volume). Scuba diving has the potential to support the maintenance and protection of the marine environment: first because the diving centre has to offer a clear and interesting underwater area to the divers; and second, because, for similar reasons, the centre will want an abundance of marine life. In this case, the diving centre may even act as a 'guardian' against improper or unauthorised fishing practices.

There are a number of examples where diving centres, either locally or through an excursion, have collaborated with local authorities to explore, clean and protect underwater areas. Diving groups regularly organise coast and seabed cleaning activities. In one such case, in Agios Nikolaos, Crete, the diving centre organised an excursion and the divers cleaned a small sea lake of remains of Second World War equipment and old military equipment. The removed items were sufficient to suggest the creation of a local exhibit for the benefit of tourists (Argyrakopoulos, 2001).

Thus, the continuing development of scuba diving as a tourist activity can be considered within the sustainable activities of marine ecotourism.

Future potential and prospects

It seems that the important underlying issue for further development is the opening up of more areas for pleasure diving, since the services offered are quite adequate. This is supported by the survey results.

The general characteristics of the activity in other countries suggest that this can be a 'new' form of tourism for Greece, even if it supports mostly existing coastal destinations. It can, however, help towards achieving the goal of extending of the tourist season, since scuba diving is not necessarily restricted to good weather conditions.

In research on diver satisfaction (Musa, this volume) it is reported that the diving tourist's length of stay is longer by 1.2 days than the length of stay of a typical tourist. This can be capitalised upon, particularly in relation to the extension of the tourist season. However, the marketing and promotional activities of the sport should be approached with caution, in order to prevent overcrowding and to maintain sustainability (see Wilson, this volume).

Since government policy is for the improvement of the tourist product, and since coastal tourism is and will continue to be Greece's main focus, it is hoped that scuba-diving tourism will become one of the future attractions.

The author discussed these issues with Rear Admiral Angelos Argyrakopoulos, Ret., of the Hellenic Merchant Marine, who was responsible for the coordination and the execution of the revision of the legal framework in 1997 and is probably one of the most experienced people in Greece as regards the implementation of the regulations. A selection from the persons contacted on the survey were also interviewed. From these contacts, three areas should be considered as priorities and require further work:

- Scuba diving as a pleasure activity should be officially included in the tourism products policy of the governmental organisations and the private sector's initiatives should be supported.
- The Hellenic National Tourism Organisation, as the official entity for the promotion of Greek tourism, should designate scuba diving as one of the alternative forms of tourism activity in Greece and give it the status and promotion given to other similar such activities.
- With the collaboration of the competent authorities, more locations should be freed up for scuba diving and areas designated as 'underwater parks' with the appropriate interventions (sign posting, positioning of wrecks, restriction of fishing, etc.) to allow them to develop for an underwater pleasure visit.

In conclusion, it can be argued that scuba diving is a positive attribute for marine ecotourism, as long as it is properly regulated and the activity is kept within the limits of the carrying capacity of each particular area. In this sense, given the experience of underwater scuba-diving activities in other countries, there is no reason why Greece cannot formally develop this form of tourism and benefit both from the increased tourism flows and from the higher spending that is associated with pleasure diving.

Acknowledgements

The author is grateful to the following for allowing themselves to be interviewed in the course of the research for this chapter: Rear Admiral Angelos Argyrakopoulos, Ret. of the Hellenic Merchant Marine, Mr P. Anagnostou, Chairman of Hellenic Federation of Associations of Underwater Activities, Sport Fishing and

Swimming Techniques, organisation officials, staff of Scuba-diving Centres and individual diving instructors and scuba divers

The author is also grateful to the participants in a questionnaire and telephone survey undertaken of scuba training and diving centres in Greece and to the Greek Ministry of Merchant Marine for providing the sampling frame.

References

Argyrakopoulos, A.T. (2000) *Thalatta Thalatta: Collection of Laws for Marine and Sea Activities: Part D – Underwater Activities*. Athens: Argyrakopoulos.

Argyrakopoulos, A.T. (2001) Underwater activities for leisure. *Greek Diver Magazine*, April-May, 12–19.

Anon. (1993) The de-regulated areas for pleasure dives. *Vithos Magazine* 5 (Jun.), 18–19.

Anon. (1998) Diver holiday directory/Diver classified holidays. *Diver Magazine, Britain's Diving Magazine*, June, 100–1/108–11.

Efstratiou, G. (2000) Underwater activities in our country. *Katadysi Magazine* 113 (Jan.–Feb.), 12–13.

Hellenic National Statistics Service (2001) Statistics of the Sector of Commerce and Services. (Online) [http://www.statistics.gr/gr/data/index.htm] (accessed 5 November 2001).

Katsadorakis, G. (1999) *The Natural Heritage of Greece*. Athens: World Wildlife Fund.

Pantis, J., Togridou, A. and Katselidis, K. (2001) Ecotourism in national parks: The case of the National Marine Park of Zakynthos. *Proceedings of the Preparatory Conference for the International Year of Ecotourism 2002: The Development of Ecotourism: The International Experience and the Case of Greece*. Athens: Hellenic National Tourist Organization.

Papasimakopoulos, N. (2000) Proposal on the development of diving tourism in Greece. *Proceedings of the First International Scientific Conference 'Tourism on Islands and Specific Destinations'*. Chios: University of the Aegean Business School.

Petreas, C. (2001) Scuba-diving: Can it help enhance the coastal tourism product in Greece? Paper presented at the ATLAS 10th Anniversary Conference, Tourism, Innovation and Regional Development, Dublin October 2001. European Association for Leisure and Tourism Education ATLAS.

Research Institute for Tourism (2001) *Hellenic Economy and Tourism*. Quarterly Report 11 (May). Athens: Research Institute for Tourism.

Townsend, C. (2001) The effects of environmental education on the behaviour of scuba divers: A case from the British Virgin Islands. Paper presented at the ATLAS 10th Anniversary Conference, Tourism, Innovation and Regional Development, Dublin October 2001. European Association for Leisure and Tourism Education ATLAS.

Tsekouras, G.T. and Partners (1991) The change of the mass tourism model: New forms of tourism. Athens: Consultancy Report for the Hellenic Bank of Industrial Development.

Ververis, D. (2001) Critical moment for the future of autonomous diving: The presentation of the Ministry of Merchant Marine. *Greek Diver Magazine* December 2001-January 2002, 10–14.

Wilson J.C., Garrod B. and Bruce D.M. (2001) Planning policy issues and opportunities for genuinely sustainable ecotourism: Experiences from the EU Atlantic Area. Paper presented at the ATLAS 10th Anniversary Conference, Tourism, Innovation and Regional Development, Dublin October 2001. European Association for Leisure and Tourism Education ATLAS.

Chapter 15

Marine Ecotourism in New Zealand: An Overview of the Industry and its Management

MARK B. ORAMS

Introduction

Marine ecotourism in New Zealand exists within the wider context of the nation's natural resources, the development of its tourism industry and its relatively short history of human habitation. This chapter will initially, therefore, provide a brief overview of the history of New Zealand tourism and consider the influence of the natural environment – including coastal and marine environments. It will also review the limited data available on ecotourism in New Zealand, in order to develop an understanding of the size and nature of the industry, both generally and more specifically, in coastal and marine settings. Finally, it will consider the current management regime and outline the importance and potential of the new Oceans Policy initiative currently being developed by the New Zealand government.

New Zealand's Natural Environment and Tourism

New Zealand has an international reputation as a tourism destination that is heavily influenced by images drawn from its natural environment. This reputation is utilised, and perhaps enhanced, by marketing campaigns promoting New Zealand as a 'clean and green' country (Hall, 1997) with a wondrous variety of '100% pure' natural experiences (Tourism New Zealand, 2001). As a consequence, many in the international tourism industry could be forgiven for considering New Zealand to have a highly developed and carefully managed ecotourism industry. To borrow the marketing catchphrase currently used by Tourism New Zealand to promote the country internationally, this view is not '100%' correct.

New Zealand is an isolated, relatively small group of islands with a short history of human habitation. In addition, it hosts a relatively small population (currently 3.6 million), has a diverse array of dramatic landscapes, a temperate maritime climate and a recent geological history including areas of substantial geothermal activity. It is these features, rather than any significant widespread commitment to conservation ideals, which has provided the basis for the nature-based tourism industry in New Zealand. In addition, while some enlightened management does exist and true ecotourism does without a doubt occur, much of New Zealand's tourism simply exploits natural attractions rather than contributes to them.

New Zealand's coastal and marine environment is significant not only in terms of its importance for tourism. With a coastline in excess of 15,000 km and a diverse range of ecosystems (ranging from sub-Antarctic Islands to the sub-tropics), it contains between a third and three-quarters of all native species in New Zealand (Oceans Policy Working Group, 2002a). New Zealand's exclusive economic zone (EEZ) covers around 300,000 km^2 (about 15 times the country's land area) which is reputedly the fourth largest Exclusive Economic Zone (EEZ) in the world (Ocean Policy Working Group, 2002a). Thus, the marine environment is extremely important for New Zealand, in both biological and economic terms.

Marine ecotourism in New Zealand exists, therefore, within this wider context of the natural environment and humankind's influence on it. Despite the nation's relatively short human history, tourism has been one of those significant human influences.

A Brief Overview of the History of Tourism in New Zealand

The establishment of New Zealand as a nation (in a modern European context) is relatively recent (1840). The potential importance of tourism to the development of New Zealand was, however, recognised early and the New Zealand government established a Department of Tourist and Health Resorts in 1901 (Collier, 1999). Early tourism to New Zealand focused on unique natural attractions such as the geothermal activity (geysers, boiling mud pools, thermal baths, etc.) around Rotorua, the glowworms of the Waitomo Caves and the alpine lakes and mountain scenery of the Queenstown area (Tourism New Zealand, 2001). A significant contributor and influence was also the involvement of indigenous Maori, both as guides and as a cultural attraction (traditional activities such as carving, songs and dances) (Barnett, 2001). These attractions remain important draw-cards for tourists to New Zealand today.

As has been the case in many other 'new-world' locations, early tourism was closely linked with early settlement from, predominantly, European immigrants. In fact,

> as early as the 1840s, New Zealand was the source of curiosity and the subject of many a travel writer, journalist and novelist intrigued by the new colony ...

> [C]uriosity was piqued by these early visitors' accounts of the new colony in the Pacific and did much to encourage a fledgling tourism industry. (Tourism New Zealand, 2001: 7)

The great majority of the attractions that were written about and which provided the impetus for this tourism development were associated with natural features of outstanding beauty or unusual quality. Thus, nature-based tourism was an early and significant contributor to tourism in New Zealand. However, due to the huge distance of New Zealand from most potential inbound markets (and, more significantly, the long travel time via ship) tourism numbers were small. The first official count of international tourism arrivals was made in 1903 when 5233 visitors (2726 from Australia and 1795 from the United Kingdom) were recorded (Tourism New Zealand, 2001).

Government involvement in the development of the tourism industry in the early part of last century was significant. The Department of Tourist and Health Resorts took over control of numerous resorts and parks, initiated infrastructure improvements and even became responsible for the administration of entire towns (such as Rotorua) where tourism was important (Collier, 1999). By 1922, just over 8000 international tourists visited New Zealand. However, growth was curtailed in the 1930s by the economic depression and virtually halted in the 1940s by the Second World War. Recovery occurred by the late 1940s and almost 11,000 international visitors were recorded for 1949/50. The introduction of 'tourist class airfares' provided the impetus for massive growth in New Zealand tourism during late 1950s and early 1960s so that by 1963/64 around 100,000 international tourists visited the country (Collier, 1999). From that time onwards, the growth of New Zealand as a destination has mirrored the expansion of the international airline industry, with tourism numbers growing exponentially as larger, faster and more economic aircraft were developed. New Zealand's reputation as a 'natural wonder' continued to grow and publicity from the New Zealand Government, from the travel industry and, presumably, through 'word of mouth', contributed to growth that resulted in one million international arrivals in 1992. Growth continues and recent records show 1.8 million visitors in 2001 (Tourism New Zealand, 2001).

New Zealand's development as a tourism destination has, without doubt, been strongly influenced by its nature-based attractions. While the industry has diversified significantly, particularly over the last decade, natural attractions and nature-based activities remain important and continue to be the dominant feature of New Zealand's tourism industry.

Ecotourism in New Zealand

While many of New Zealand's longstanding tourism attractions and operations have been nature based (for example, the Franz Joseph and Fox Glaciers of the South Island), 'ecotourism' operations have been relatively recent entrants into the tourism market place (Tourism New Zealand, 2001). Recent research by Higham *et al.* (2002) found that the majority of New Zealand ecotourism operations were less than 10 years old. This study is the first comprehensive overview of ecotourism in New Zealand and provides valuable insights into the industry. The study identified some 400 tourism operators for whom nature-based attractions formed a significant component of their business. However, when the more stringent criteria associated with 'ecotourism' were applied, the authors concluded that there were 247 businesses in New Zealand that could be considered ecotourism operations.

It is important, therefore, to differentiate between 'true ecotourism' in considering the industry in New Zealand from those operators who simply base their businesses on natural attractions. True ecotourism, as it is generally defined in the literature (see Weaver, 2001), is not only based on nature but is learning centred (i.e. it deliberately incorporates an educational component) and is managed in a way that attempts to be environmentally, socio-culturally and economically sustainable. What is interesting is that when these more stringent criteria were applied in the Higham *et al.* study, it was found that over half of New Zealand's nature-based tourism businesses could be considered to be ecotourism operators (at least in intent). The study concluded that ecotourism operations in New Zealand were diverse and ranged from small-scale, part-time 'life-style' businesses to attractions managed via non-profit organisations, government agencies or large corporations that hosted visitor numbers in excess of 100,000 per annum.

Marine ecotourism in New Zealand

A census of marine tourism operators in the mid-1990s by McKegg *et al.*, 1998) identified 376 businesses that had some type of marine attraction as part of their operation. Marine tourism was defined for the purposes of this study as 'commercial operations visiting natural areas for the purpose of diving, fishing, marine mammal and seabirdwatching, cruising and tour boating' (McKegg *et al.*, 1998: 154). The research profiled the industry and found, not surprisingly, a strong seasonal bias in client numbers to the summer months (December to March). In addition, their findings are consistent with the more recent Higham *et al.* (2002) study in that they conclude that the marine tourism industry is relatively new in New Zealand and growing quickly – with over 60% of operations being established in the past five years. It is also interesting to note that marine tourism operators were found to be predominantly small, with the great majority (82%) comprising 'local' operators employing less than three people. As a result, McKegg *et al.* (1998) conclude that many marine tourism operations in New Zealand struggle to be commercially viable year round.

Interestingly, McKegg *et al.*'s (1998) work identified activities that are typically considered in the 'ecotourism' realm as the most popular marine tourism activities in New Zealand. They found wildlife viewing to be the most commonly mentioned attraction offered by businesses, with 60% of operators noting seabirds (including penguins) and 44% noting marine mammals as a 'key attraction' of their tour. Of this wildlife, dolphins were the species (four species are commonly found in New Zealand) most frequently focused upon (22% of operators) followed by New Zealand Fur Seals (one species) and penguins (three species). Additional activities

Table 15.1 Attractions targeted by New Zealand marine tourism operators

Attraction	*Percentage of operators offering attraction*
General scenery	52
Marine mammals	44
Sea birds (excluding penguins)	42
Fish (including fishing)	30
Islands	21
Penguins	18
Other marine wildlife	16
Marine reserves	5
Marine farming	5
Special habitats	4
Crustaceans	4
Shellfish	4
Shipwrecks	3

Source: Derived from McKegg *et al.* (1998).

found to be significant were grouped in a category termed 'scenery' – this included the viewing of attractions associated with the coastal and marine environment such as caves, historic sites and 'general marine vista' (McKegg *et al.*, 1998: 155). Table 15.1 shows the range of activities (ranked in order of frequency) targeted by New Zealand marine tourism operators.

What is not explicitly clear from the work of McKegg *et al.* is how many of these marine tourism operations could be considered ecotourism. It appears a safe assumption that the great majority could be considered ecotourism (general scenery, marine mammal and seabirdwatching and visiting marine reserves); however, many activities such as fishing, shellfish and crustacean gathering and marine farming are likely to fall outside of the ecotourism realm.

This presents a major challenge in attempting to quantify marine ecotourism operations in New Zealand (and probably elsewhere). Marine tourism, because of its focus on the sea or coast, is, by definition, 'nature-based' (Orams, 1999); however, it is contentious as to whether or not specific marine tourism businesses are actually 'ecotourism' operations. For example, despite the winning of 'ecotourism' awards and becoming one of the promotional 'flagships' for ecotourism in New Zealand (Tourism New Zealand, 2001), some have questioned whether Whale Watch Kaikoura (with up to 100 trips per week focused on usually fewer than 10 whales) is environmentally sustainable. Whether the operation offers educational programmes of sufficient standard to qualify as 'ecotourism' has also been the subject of debate (Orams, 2002).

In a similar way, tourist trips to islands in the Hauraki Gulf Marine Park differ significantly in terms of their character, purpose and impacts. Tiritiri Matangi island is a location where ecotourists have, through planting programmes and financial support, significantly improved the quality of the island's flora and fauna (Orams, 2001). However, nearby Rangitoto island features a commercial tourism concessionaire who provides a 'motorised train ride' up a sealed road, for visitors to access the view point at the summit of the volcano (personal observation). Thus, even though both islands are managed by the Department of Conservation for conservation purposes, the character (and impact) of the tourism is entirely different. This illustrates how difficult it is, from a generic study of marine tourism or from an analysis of databases on tourism operators, to judge accurately whether or not businesses are actually ecotourism operations.

The study by Higham *et al.* (2002) does, however, provide a reasonably accurate impression of the importance of the coastal and marine environment for ecotourism. This is because it examined ecotourism specifically, as opposed to the McKegg *et al.* study which focused on marine tourism generally. Of the 247 ecotourism operations identified by Higham *et al.* in New Zealand, 116 (47%) can be identified as solely marine or coast based. In addition, Higham (pers. comm. 1 March 2002) also notes that many of the remaining operations were 'general' nature tours, many of which have a secondary focus, if not a primary focus, on the coastal-marine environment. This finding is consistent with the conclusions of the 'Oceans Policy Working Group', a team working on the development of an 'Oceans Policy' for the New Zealand government (see www.oceans.govt.nz) who state that there is:

> A strong relationship between tourism and the sea in New Zealand – the coastal backdrop, the harbours, beaches to walk on and the sounds and smells of the ocean.
> (Oceans Policy Working Group, 2002b: 10)

They also note that around 76,000 tourists (6% of all international visitors) go recreational fishing each year. In addition, tourists' activities mentioned (unprompted) in the annual New Zealand International Visitor Survey with relevance to marine tourism were 'scenic cruise' (334,000 or 19% of all international visitors) and 'visited beaches' (345,000 or 19% of all international visitors) (Tourism New Zealand, 2002). Whether such activities actually constitute 'ecotourism' is not clear, however, it provides additional evidence regarding the importance of the marine environment for tourism in New Zealand.

The finding that New Zealand's marine ecotourism industry focuses very much on marine wildlife and scenery is not surprising, given that many other tourism destinations in the South Pacific offer tropical marine settings (which New Zealand does not have) that are often more attractive to tourists (such as coral reefs). A number of species that form the 'mainstay' of New Zealand's marine ecotourism 'product' are unique to New Zealand. Examples include Hector's dolphins, New Zealand (Hooker's) sea lions and Hioho (yellow-eyed) penguins. Others offer a 'competitive advantage' for New Zealand, because opportunities for similar experiences (species) do not exist at alternative destinations close by (such as Australia or the South Pacific islands). Examples include sperm and Bryde's whales, royal albatrosses and dusky and common dolphins. In addition, New Zealand possesses accessible coastal scenery – such as deep-water fiords and active volcanic islands – which is unique in the South Pacific. Thus, New Zealand's marine ecotourism industry has tended to focus on those attractions which are readily accessible and which offer some competitive advantage over alternative destinations. This is particularly important, of course, for the Australian market, which forms New Zealand's single largest inbound group (35% of all international arrivals) (Tourism New Zealand, 2002).

While there is no research specifically directed at the marine ecotourism industry in New Zealand, information that is available (such as that reported earlier) does provide some clear indications regarding the size, characteristics and importance of the industry. What is clear from these studies is that marine ecotourism is a dominant feature of the New Zealand ecotourism scene. This is perhaps not surprising given that the country is an island nation with a long and intricate coastline and a relatively temperate climate. However, it represents a major transition from New Zealand's early tourism development, where alpine and geothermal attractions dominated (Collier, 1999; Tourism New Zealand, 2001).

From the limited information available, several general conclusions about marine ecotourism in New Zealand can be made.

(1) While marine attractions have a long history as important components of the tourism industry in New Zealand, marine ecotourism is a recent entrant into the New Zealand tourism scene.
(2) Growth has been rapid, with over half of all marine ecotourism businesses being established in the last decade.
(3) Marine ecotourism operations are diverse, ranging from large corporations hosting hundreds of thousands of tourists each year (with annual turnovers of tens of millions of New Zealand dollars), to small owner/operator businesses that struggle to survive financially. However, the industry is predominantly made up of small operators who employ less than three people. Examples of this diversity include 'Whale Watch Kaikoura' a well-known and financially-successful large corporation (Orams, 2002). At the other end of the scale are operations like 'Mercury Bay Seafaris', a husband and wife operation that only operates over the summer, hosts hundreds of visitors a year and barely makes enough money to cover costs (Ryan, 1998).
(4) Patronage of marine ecotourism businesses is highly seasonal and many operators struggle to remain financially viable over the winter months.
(5) High quality and unique coastal environments are important aspects of the overall 'scenery' for marine ecotourism trips.
(6) Marine wildlife – particularly dolphins, seals, penguins and other sea birds – are particularly important attractions – so much so that now coastal and marine settings are the most significant ecotourism environments (at least in terms of operators) in New Zealand.
(7) A significant proportion of marine tourism businesses can be considered to be

ecotourism (in theory). However, there remains considerable 'blurring' between ecotourism as a concept and the application of the concept in marine tourism scenarios.

Management of ecotourism

The management of New Zealand's natural resources, and the utilisation of them for activities such as tourism, is governed by a complex series of laws and a multiplicity of government agencies. While it has been simplified in recent decades, challenges and, inevitably, frustrations arise – particularly for organisations such as marine tourism operators, who commonly move from one jurisdiction to another in the course of their activities (for example, as when crossing the mean high water mark). These kinds of jurisdictional 'overlap' and the complex nature of management regimes dealing with marine resources is typical and certainly not unique to New Zealand (Kenchington, 1990). However, a more fundamental issue exists for New Zealand that has relevance to the management of ecotourism. At a national and international level, two major national governmental agencies have important roles and influence over the tourism industry, Tourism New Zealand (Tourism NZ) and the Department of Conservation (DoC). These two agencies have, as Simpson (in press) points out, roles which are potentially conflicting rather than complementary.

Tourism New Zealand

Tourism NZ was created by the New Zealand Tourism Board Act 1991 and it exists primarily to 'ensure that New Zealand is developed and marketed as a competitive tourism destination to maximise long term benefits to New Zealand' (New Zealand Tourism Board Act, 1991: section 6). This has, understandably, been interpreted by Tourism NZ so as to view their role as a promotion and marketing agency, charged with attracting ever-increasing numbers of international tourists to the country – preferably spending ever-increasing amounts during their visits. Since its formation, Tourism NZ has been successful in facilitating growth in international visitor arrivals that is, overall, in excess of most developed nations (New Zealand's international arrivals have almost doubled from one to nearly two million in the past decade). A major feature of the promotional strategies of Tourism NZ has been to emphasise the high-quality natural attractions (including those of a marine nature) of the country. In addition, Tourism NZ (and the tourism industry itself) has been quick to embrace the label of ecotourism from a marketing perspective. However, Tourism NZ has not taken on a role in the actual development of sound ecotourism management practices in New Zealand. There are currently no government-based or government-supported ecotourism training programmes, no accreditation of operators and no ongoing systematic government assessment of the impacts of ecotourism in New Zealand. In addition, the tourism industry itself has developed a National Tourism Strategy (released May 2001) that appears similarly biased toward increasing visitor numbers, without any concrete proposals about how the impacts of these increasing numbers should be managed (particularly in an ecotourism context). Surprisingly, there are few voices expressing concern over this 'more is better' approach to tourism in New Zealand. An exception is an important recent contribution from Simpson (in press) who states:

> It often appears that the tourism industry, including some so-called ecotourism operators, continue to regard environmentalism as a necessary evil, something that is promotionally useful but operationally restrictive; at the same time, that industry continues to exploit clean, green and unspoiled imagery in its external promotion, threatening to provide a text book example of killing the goose that laid the golden egg. (Simpson, in press)

However, concerns over the impacts and lack of management of ecotourism in New Zealand have been expressed as far back as the early 1990s by Warren and Taylor (1994) and Hall (1994) and, more recently by Ryan (1998), Pearson (1998) and Dowling (2001) who concluded:

> In New Zealand, ecotourism is at a relatively embryonic stage without an agreed definition, formalized national strategy or accreditation scheme, its ecotourism development faces some challenges. This is particularly so because the demand for New Zealand nature-based tourist activities is high, and consequently the industry is demand-led. This is a situation that could have significant impacts on the resources if they are not managed appropriately. (Dowling, 2001: 146)

The Department of Conservation

There is no specific governmental entity charged with the responsibility of managing ecotourism in New Zealand (there is a Ministry of Tourism but it lies under the Department of Commerce and is primarily a small group of policy advisers). However, in terms of the supply and management of ecotourism attractions, the most important government agency is the Department of Conservation (DoC). Established in 1987 by the Conservation Act, the DoC is the national government agency charged with the conservation and sustainable use of 'Crown' (public) lands, including national parks and reserves and, significantly, all foreshore (below mean high water) and seabed within the coastal and marine area. In addition, the DoC is charged with management of marine reserves, historic places, endangered species and marine mammals. While DoC has conservation as its primary management role, it has also recognised the important role it plays in providing for recreational and tourism opportunities. This is important because 'a substantial majority of outdoors tourism takes place on public lands under DoC control' (Simpson, in press).

With the exception of three small, locally managed marine protected areas, the DoC has direct responsibility for the management of all of New Zealand's marine reserves. There are currently 16 marine reserves and two marine mammal protected areas, totalling around one million hectares of sea under protection. This constitutes approximately 5% of New Zealand's 'territorial sea' (mean high water mark to 12 nautical miles offshore). However, three-quarters of this area is contained in one offshore marine reserve at the remote, uninhabited, Kermadec Islands. For 'mainland' New Zealand, only around 15,000 hectares or 0.1% of the territorial sea is under protection (Oecans Policy Working Group, 2002c). Campaigns to increase the number of marine reserves (protected areas) in New Zealand significantly have been longstanding but largely ignored by successive governments (Ballantine, 1991).

The DoC also has an important strategic planning role in that it establishes the New Zealand Coastal Policy Statement (as required under the Resource Management Act 1991) and is responsible for the approval of restricted coastal activities and regional coastal plans. The management of tourism is primarily considered in the Department's Visitor Strategy (DoC, 1996). This strategy is underpinned by five fundamental goals relating to environmental protection (clearly the first priority), recreational visits, tourism activity, information/education and visitor safety. What is challenging for the DoC is that its mandate sets a priority for environmental conservation, whilst allowing for appropriate recreation and yet it is faced with ever increasing numbers of visitors (both of international and domestic origin) to the attractions it is charged with protecting. This classic resource management dilemma is not unique to the DoC. However, it is complicated by an increasing expectation that the DoC will generate income from tourist uses of the 'national estate'. DoC manages these commercial tourist uses of public resources via a series of commercial 'concessions' (licences) and the granting and administration of these concessions has become a growing management challenge for the department.

Additional management agencies

Other important management agencies include the Maritime Safety Authority, who administer regulations controlling maritime transport (including the survey and registration of commercial vessels used for transporting passengers), and the Ministry for the Environment, who provide an overview of environmental policy for the government, including setting procedures for environmental impact assessment. The Ministry of Fisheries administers laws and regulations designed to ensure the sustainable use of New Zealand's fish resources (including shellfish and crustaceans). As part of this role, they set catch limits and minimum sizes for many species popular with recreational fishers. Regional Councils are responsible for the creation and implementation of Regional Coastal Plans. These plans may include restrictions over certain activities and/or the requirement of a formal consent for particular developments, activities or in specified areas. Thus, any activity (including tourism) likely to have a significant environmental impact (or that requires a structure, such as a wharf) requires a 'resource consent' from the relevant regional council. Furthermore, Regional Councils are responsible for the creation and implementation of navigation and safety bylaws and maritime rules. At a local level, individual territorial local authorities (city and district councils) control land use above the mean high water mark (within their jurisdiction). Consequently, marine operations that wish to establish shore-based facilities (such as offices, advertising, pick-up and drop-off points, and so on) must do so under the management of the local council's 'district plan', rules and bylaws.

The role of indigenous Maori

An additional issue of importance in New Zealand's management of its marine resources (including their utilisation for tourism) is the rights of Maori. Under the Treaty of Waitangi (the agreement between indigenous Maori and European immigrants signed in 1840) and subsequent legislation, Maori are guaranteed a number of rights relating to traditional uses of the sea as a source of food and as an area of spiritual and cultural significance. Important concepts dealt with under the Treaty include the rights of Maori to have *rangatiratanga* (control or sovereignty) and *kaitiakitanga* (guardianship) over their *taonga* (treasures), including their natural, cultural and spiritual resources. As a result, laws (such as the Resource Management Act 1991), make provision for the specific inclusion of Maori and Maori values in decision-making, planning and management of natural resources (including the sea). Examples of this involvement include the management of customary fishing areas (such as Taiapure and Mataiai) and roles as *kaitiaki* (caretakers or guardians) of coastal and marine areas.

Maori have always played a significant role in the New Zealand tourism industry (Barnett, 2001). However, there has also been a recent legal recognition of the important role Maori play in the ecotourism industry. In a high-profile court battle, the Ngai Tahu *iwi* (tribal grouping) challenged the DoC's plans to grant permits for additional operators to offer whalewatching tours out of Kaikoura – a small town of 10,000 on the South Island's east coast. Significantly, Ngai Tahu were the owners of the only existing whalewatch business in the area. They argued that whales were and are culturally significant for Maori and are '*taonga*' (cultural treasures). More specifically, Ngai Tahu claimed that they had, and continue to have, '*rangatiratanga*' over the whales as a resource at Kaikoura. Consequently, according to Ngai Tahu, whales and the use of them (even in a modern context) are covered under the principles of the Treaty of Waitangi and, as a result, that the use of whales for any purpose (including ecotourism) was something that Maori had specific (and exclusive?) rights toward. In effect, such a court challenge was unique in that it attempted to protect a commercial (and lucrative) monopoly on the basis of cultural rights. This issue was taken via New Zealand's High Court to the Court of Appeal in 1994 and 1995 where the Court's finding provided some support for Ngai Tahu's position without entirely agreeing that the 160 year Treaty of Waitangi could have foreseen the use of whales as a tourism resource. What the Court did state was that rights pertaining to economic development for indigenous peoples were becoming 'recognised and accepted in international jurisprudence' and that the principles of the Treaty of Waitangi required 'active protection' of Maori interests (Orams, 2002). As a consequence, the court directed the DoC to take into account the protection of Ngai Tahu interests before awarding further whalewatching permits. The end result is that no further permits have been issued for whalewatching in

Kaikoura and the Ngai Tahu-owned Whale Watch Kaikoura remains the sole operator in the area.

This finding has significant implications for the management of marine ecotourism in New Zealand. It provides strong support for the argument that Maori's interest and value in marine resources – including their utilisation in a modern commercial context as tourism attractions – should be given due and careful consideration in management decisions. While there has not, as yet, been additional legal attempts to protect commercial (and cultural) interests for Maori in a marine ecotourism situation, this is sure to arise as marine ecotourism continues to grow.

While the special status of Maori in marine resource management and planning is appropriate and consistent with the philosophy of ecotourism, in practice it has proved frustrating for many in New Zealand. The challenges in affording Maori special status (as required under the Treaty of Waitangi and resource management law) is provided by a decision making process grounded in western democratic principles (such as majority decision making) whilst Maori culture provides a different context for decision making (consensus). This is complicated further by the fact that Maori themselves seldom have a consistent opinion on resource management issues and, inevitably, difficulties in incorporating ('valuing') intangible issues, such as spiritual significance, arise. In addition, the legal requirement to incorporate Maori in planning and decision making has placed significant demands on the Maori population. In particular, Maori leaders and Maori who are educated in, or familiar with, resource management laws, policies and procedures are in high demand. This has resulted in 'burn-out' for some, whilst a few have taken the opportunity to charge for their involvement in decision-making processes.

The end result has been, in some cases, a 'backlash' and resentment of Maori and their special status in New Zealand resource management; but in others meaningful and worthwhile involvement of Maori which has enhanced the decision-making process and resulted in better outcomes for marine resources. One of the important challenges for the future in managing New Zealand's marine resources (including their use for tourism) therefore, is to continue to strive to achieve partnership with Maori and to give meaningful effect to the principles of the Treaty of Waitangi.

Management Challenges for Marine Ecotourism in New Zealand

In New Zealand, there is a government agency (Tourism NZ) charged with promoting New Zealand internationally as a tourism destination and delivering ever increasing numbers and expenditure from visitors to the country. Naturally enough, this agency heavily utilises imagery from the natural environment and enthusiastically adopts the positive connotations associated with promoting New Zealand as a kind of ecotourism world leader. On the supply side, the DoC continues to try and accommodate visitation to many of these natural attractions, without having that use degrade the resource. Thus, the concept of (true) ecotourism is also an attractive one for the DoC. Similarly, other government agencies – while having a mostly peripheral role in ecotourism management – also have a legal requirement under New Zealand's Resource Management Act (1991) to ensure that the effects of development and activities (including tourism) are environmentally sustainable.

What is critically missing in the management of New Zealand's important tourism industry is any practical vision, strategy and system for actually implementing ecotourism on the ground (and water). New Zealand lags far behind Australia in this regard. Australia has an established and well-supported National Ecotourism Strategy (established in 1994). It also has an active national ecotourism association, formed in 1991 (and local versions), the publication of an annual industry guide, a code of ethics, an accreditation programme, and a training and research agenda (for example, their Cooperative Research Centre for Sustainable Tourism). It is shameful that New Zealand has not shown a similar commitment to actually implementing and managing ecotourism in a real sense.

The Oceans Policy Initiative

It is well recognised that New Zealand's coastal and marine environments have a complex and overlapping (and often confusing) set of management regimes and agencies with varying responsibilities. This often results in (or at the very least contributes to) the frustration that many marine tourism operators experience in both establishing and managing their businesses. In addition, it is highly likely that this complexity enhances the risk of negative impacts and contributes to difficulties and conflict amongst those involved in use of the coastal-marine environment. In briefing documents, the Oceans Policy Working Group (2002d) state:

> There is a lack of communication and a grave lack of trust among major stakeholders. This is severely inhibiting the advancement of sustainable management of the marine environment. (Oceans Policy Working Group, 2002d: 1)

The New Zealand government has recently taken action to try and improve this situation, because they recognise:

> Sustainability is threatened by the lack of an overarching management framework or strategy that guides positive collaboration between stakeholders, and integrates diverse interests and values into environmental management solutions. (Oceans Policy Working Group, 2002d: 1).

As a consequence, a major initiative has been undertaken to create a New Zealand Oceans Policy. This two-year (initially) process involves extensive public consultation and the development of proposals to the New Zealand Government to address the concerns of New Zealanders with regard to future utilisation and health of the nation's marine environment. The outcome of this process is difficult to predict but is likely to include a series of guiding principles regarding the wise use and management of the sea and recommendations regarding a legal framework which could simplify the current system to render it more effective and less complex. It is hoped (by this author at least) that the management of marine ecotourism in New Zealand could be rendered more effective and successful by these changes.

Conclusions and Important Issues for the Future

New Zealand has a relatively recent history of human habitation and, furthermore, the density of human population is, in world terms, low. In addition, the country is blessed with a diverse array of dramatic landscapes, a temperate maritime climate and an intricate and varied coastline that includes an abundance of plant and animal species and habitats. Unfortunately, however, the influence of humans on New Zealand's natural environment has been massive. New Zealand has an unenviable record of loss of species, deforestation and resource exploitation. While most of this effect has been on terrestrial environments, the marine environment has not been immune and numerous examples exist of over-exploitation and a resulting species population collapse. No case is more illustrative than that of New Zealand's exploitation of great whales. During the 1950s and early 1960s, New Zealand hosted several commercial whaling stations. These stations primarily targeted humpback whales, killing some 10,000 during their brief operation. The industry closed down because there were so few whales left that it was no longer economically viable. Today, some 40 years later, the humpback whales that were once so abundant along New Zealand's coast during their winter migrations are seldom found.

Similarly, other human activities have been, and are still, detrimentally affecting New Zealand's marine environment. Destruction of wetlands, dredging, dumping of sewage effluent and 'stormwater', over-fishing, urbanisation of catchments, marine debris and litter, the use of pesticides and herbicides, land 'reclamation' and the introduction of exotic species are all features of New Zealand's 'use' of its marine environment. This degradation has, of course, significant consequences for New Zealand's ecotourism industry; it is prerequisite that an accessible healthy, clean and abundant marine environment exists for the ecotourism industry to survive and prosper. Thus, the future health of the seas surrounding New Zealand is an issue of fundamental importance for New Zealand tourism.

New Zealand has ambitious targets to continue to increase international tourist arrivals over the next decade. In addition, an increasing number of New Zealanders wish to use the sea for recreational purposes. Recreational use of New Zealand's marine environment is, therefore, going to continue to increase and have a major influence on life in and surrounding the country. Over the next 50 years, curiosity for new and unexplored environment, combined with increased technological capability, will see increasing numbers of New Zealanders and tourists visit many new locations previously 'protected' from direct human influence. As a consequence, both the demand for and impact on high-quality marine environments will increase.

While the rapidly growing use of New Zealand's seas for recreation and tourism provides significant challenges, it also provides a great opportunity. Ecotourism, by definition, seeks to make a positive contribution to the welfare and health of the ecosystems which act as its host. Ecotourists can, through visiting and enjoying the marine environment, provide an increased impetus and economic incentive for marine conservation. One of the most important considerations and opportunities for the new Ocean's Policy (and marine management regime) for New Zealand is to ensure that the demand for the use of marine resources for recreational purposes (including tourism) is given sufficient precedence in decision making about marine resource utilisation.

Acknowledgements

The author acknowledges the variety of sources and influences on the material presented in this chapter. In particular, the works of Alan Collier, Ken Simpson, James Higham, Anna Carr, Stephanie Gale and Susan McKegg have been helpful in writing this chapter. The comments and suggestions of the editors of this volume have also been helpful.

References

Ballantine, W.J. (1991) Marine reserves for New Zealand. *Leigh Laboratory Bulletin* 25. Auckland: University of Auckland.

Barnett, S.J. (2001) Maori tourism. In C. Ryan and S. Page (eds) *Tourism Management. Towards the New Millennium* (pp. 446–50). Oxford: Elsevier Science.

Collier, A. (1999) *Principles of Tourism. A New Zealand Perspective* (5th edn). Auckland: Addison Wesley Longman.

Department of Conservation (1996) *Visitor Strategy.* Wellington: New Zealand Government Printer.

Dowling, R.K. (2001) Oceania (Australia, New Zealand and South Pacific). In D.B. Weaver (ed.) *The Encyclopaedia of Ecotourism* (pp. 139–54). Wallingford: CABI.

Hall, C.M. (1994) Ecotourism in Australia, New Zealand and the South Pacific: Appropriate tourism or a new form of ecological imperialism? In C. Cater and G. Lowman (eds) *Ecotourism: A Sustainable Option?* (pp. 137–57). Chichester: John Wiley and Sons.

Hall, C.M. (1997) *Tourism in the Pacific Rim. Development, Impacts and Markets,* 2nd edn. Melbourne: Longman.

Higham, J.E.S., Carr, A.M. and Gale, S. (2002) Ecotourism in New Zealand: Profiling visitors to New Zealand ecotourism operations. He tauhokohoko nga whakaaturanga a nga manuhiri ki rawa whenua o Aotearoa. Research Paper Number 10, Department of Tourism, University of Otago, Dunedin.

Kenchington, R.A. (1990) *Managing Marine Environments.* New York: Taylor and Francis.

McKegg, S., Probert, K., Baird, K. and Bell, J. (1998) Marine tourism in New Zealand: A profile. In M.L. Miller and A. Auyong (eds) *Proceedings of the 1996 World Congress on Coastal and Marine Tourism,* 19–22 June 1996, Honolulu, Hawaii, USA (pp. 154–9). Washington Sea Grant Program and the School of Marine Affairs, University of Washington and Oregon Sea Grant Program. Corvallis, Oregon: Oregon State University.

New Zealand Tourism Board Act (1991) *New Zealand Tourism Board Act.* Wellington: New Zealand Government Printer.

Oceans Policy Working Group (2002a) New Zealand's marine environment: What are we dealing with? Information Sheet No. 1 (online). [www.oceans.govt.nz;] (accessed 1 March).

Oceans Policy Working Group (2002b) Uses and values of the marine environment. Information Sheet No. 2 (online). [www.oceans.govt.nz;] (accessed 1 March).

Oceans Policy Working Group (2002c) Protecting our oceans. Information Sheet No. 3 (online). [www.oceans.govt.nz;] (accessed 1 March).

Oceans Policy Working Group (2002d) How is the current system performing? Information Sheet No. 7 (online). [www.oceans.govt.nz;] (accessed 1 March).

Orams, M.B. (1999) *Marine Tourism. Development, Impacts and Management*. London: Routledge.

Orams, M.B. (2001) Types of ecotourism. In D. Weaver (ed.) *Encyclopedia of Ecotourism* (pp. 23–36). Wallingford: CABI.

Orams, M.B. (2002) Marine ecotourism as a potential agent for sustainable development in Kaikoura, New Zealand. *International Journal of Sustainable Development* 5 (3), 338–52.

Pearson, S. (1998) An ecotourism strategy for New Zealand. In J. Kandampully (ed.) *Advances in Research (Part 1) Proceedings of the Third Biennial New Zealand Tourism and Hospitality Research Conference*. Canterbury: Lincoln University.

Ryan, C. (1998). Dolphins, canoes and marae – ecotourism products in New Zealand. In E. Laws, B. Faulkner and G. Moscardo (eds) *Embracing and Managing Change in Tourism* (pp. 285–306). London: Routledge.

Simpson, K.G.M. (in press) Ecotourism policies in New Zealand. In D. Fennell and R. Dowling (eds) *Ecotourism Policy*. Wallingford: CABI.

Tourism New Zealand (2001) *100 Years Pure Progress. 1901–2001, One Hundred Years of New Zealand Tourism*. Wellington: Tourism New Zealand.

Tourism New Zealand (2002) Database at http://www.tourisminfo.govt.nz; 4 April.

Warren, J.A.N. and Taylor, N.C. (1994) *Developing Ecotourism in New Zealand*. Wellington: New Zealand Institute for Social Development and Research.

Weaver, D. (ed.) (2001) *The Encyclopaedia of Ecotourism*. Wallingford: CABI.

Conclusions

BRIAN GARROD and JULIE C. WILSON

Marine ecotourism is a new phenomenon. This statement might seem an odd one to make, given that the activities covered by this term, including watching whales and dolphins, scuba-diving and snorkelling, beach-walking and rockpooling, have clearly been established forms of recreation for several decades, if not far longer in some cases. Marine ecotourism is, however, more than simply the sum of the activities that might be considered to fall within its remit: it represents an approach to the conduct of such activities that is intended to achieve superior outcomes to those associated with more conventional formulations of tourism. This claim of superiority has often been expressed in terms of the sustainability paradigm: the ecotourism approach is oriented towards achieving a more sustainable relationship between tourism, the environment in and upon which it is based, the local economy of which it is part, and the host community in which it is set. The concept of sustainability, meanwhile, is often dated to the late 1980s, when it was first articulated in detail by the World Commission for Environment and Development in what has become known as the 'Brundtland Report'. Arguably, then, it is marine ecotourism's explicit orientation toward sustainability that makes it a new phenomenon.

Marine ecotourism is also a relatively fast-growing segment of the world tourism industry, which has itself been a major growth area in the global economy in recent years. Whalewatching, for example, is thought to be growing in value terms by around 12% per annum (see Berrow, this volume). From small beginnings in California in the mid-1950s, whalewatching has become a significant worldwide phenomenon, with whalewatching tours now taking place in 87 countries around the world, as well as the Antarctic. The result of this rapid growth in marine ecotourism has been the development of an industry that is capable of having substantial impacts – both positive and negative – on the natural environment, on local and national economies and on the social fabric of host communities. It is clear, therefore, that proponents of marine ecotourism cannot pin their hopes on achieving the sustainability goals set for marine ecotourism by keeping its activities small in size and restricted in geographical terms: the industry has already grown too large and too widespread for that to be a realistic strategy. Those promoting marine ecotourism must instead aim to achieve their goals by managing and reducing the negative impacts of its activities. Merely trying to contain such impacts by keeping marine ecotourism spatially limited and small in scale will not work. At the same time, it is increasingly being recognised that ecotourism development has, in the past, often failed to deliver the benefits it has promised in terms of support for conservation efforts, economic benefits for local people and social cohesion for the host community. For marine ecotourism to be truly sustainable those who have an interest in it must ensure that it delivers an optimum balance of positive and negative impacts. Yet it is clear that this requires a better understanding of what such impacts are, why they occur and how they might best be addressed. With this in mind, one of the major aims of this book has been to advance and disseminate knowledge in this area.

The aim set for marine ecotourism for it to become increasingly sustainable confers a number of specific obligations on those who plan and deliver it. In other words, if marine ecotourism is to lay a justifiable claim of being a more sustainable form of tourism, the activities covered by its remit must be palpably different in their orientation to activities that are not. However, one emerging realisation in the small but growing literature on ecotourism is that the industry is not especially well endowed with the knowledge and expertise required to adapt its activities according to the principles of sustainability. This is especially true of marine ecotourism: as several of the chapters in this volume have noted, there has been little research into the biology of the many, often endangered species that have become the targets of marine ecotourism operations. The same is broadly true of research into the ecology of the typically fragile ecosystems in which marine ecotourism takes place or upon which it has indirect impacts. For example, virtually nothing is known about the lifecycle of certain species of whale that are targeted by whalewatching operators: even basic data on when and where such animals mate, calf or even feed are highly imperfect. Meanwhile, proper scientific studies of the impacts of marine ecotourism activities are rare. This is true not only of studies of the environmental impacts of marine ecotourism from the natural sciences but also of studies of the impacts of marine ecotourism on economies and on local communities from the social sciences. The paucity of relevant research makes it difficult for those concerned in planning and managing marine ecotourism to do so effectively. This also creates problems when attempting to substantiate the success or otherwise of marine ecotourism in meeting the sustainability objectives that are set for it.

A broadly similar situation exists with regard to the tools available for regulating and managing marine ecotourism. While a number of such tools have been developed and implemented in recent years, studies evaluating their relative performance in meeting the objectives for marine ecotourism remain few and far between. This situation makes it very difficult indeed for those responsible for planning and managing marine ecotourism to learn from previous experiences and make progress in ensuring that the sustainability objectives for marine ecotourism are properly set and rigorously pursued. Part of the purpose of this book has been to examine a range of experiences of marine ecotourism around the world and to draw out some of the practical lessons that are there to be learned.

A further challenge for proponents of marine ecotourism is the commonly-held misconception that it simply represents a type of tourism activity, rather than an overall approach to planning and managing certain types of marine tourism activity in more sustainable ways. The result has been that a vast number of so-called 'ecotours' have sprung up all over the world. Such tours employ the image of marine ecotourism but do little or nothing to ensure that they are oriented toward the achievement of sustainability. An example cited in this book is shark-cage diving at Gansbaii, South Africa, where the term 'ecotourism' seems to be being used simply as a marketing device. Arguably, these 'sham' operators have been able to continue operating because of a lack of consensus on what constitutes the objectives of 'true' marine ecotourism, as well as the absence of scientifically validated metrics to assess the relative performance of operators in achieving them. A concern expressed in this book is that the damaging activities of operators using the ecotourism concept merely as a marketing device to capture a share of this rapidly growing tourism market may ultimately bring the concept into disrepute, spoiling the market for 'counterfeit' and 'genuine' marine ecotourism operators alike. Indeed, there is some evidence to suggest that certain operators in Australia are choosing not to describe their products as ecotourism because they feel that the term has negative connotations.

Arguably, then, marine ecotourism is presently at a critical stage in its development, both as a concept and as practical orientation. The obligations of participants in marine ecotourism must be identified and understood more clearly, with a wider agreement as to their implications and with more effective application. In short, genuine marine ecotourism needs to be able to be distinguished clearly from counterfeit versions. This requires the marine ecotourism industry to be clear about what objectives it is trying to achieve and how precisely it intends to go about achieving them. All of this will, of course, necessitate both operators and tourists working toward a common understanding of what marine ecotourism really is or, perhaps more significantly, what it really should be.

Local Participation

Most definitions of marine ecotourism emphasise the principle that local people should take an active role in decision making at each stage and in every aspect of its development. The inclusion of this principle undoubtedly reflects lessons learned from the experiences of the earliest ecotourism development projects, such as the introduction of wildlife-watching tourism in the East African game parks, which tended to be planned in a largely 'top-down' manner. Ideally, local people should not simply be beneficiaries of ecotourism: they should be participants in it, sharing in the decision making that determines what the costs and benefits of developing ecotourism are to be and how they are to be shared. Indeed, the widely acknowledged success of more recent ecotourism programmes in East Africa, such as the CAMPFIRE programme in Zimbabwe, has been ascribed largely to the degree of success achieved in empowering local people to participate in the processes of planning and managing ecotourism.

It is generally agreed, therefore, that local participation must be made a central feature of marine ecotourism if it is to be effective in meeting its overall objectives. This implies shifting the balance of power to the local level, enabling the local community not only to receive a more equitable share of the benefits of marine ecotourism but also to make a meaningful contribution to the various decision-making processes involved in planning, managing and developing marine ecotourism. Indeed, it is widely argued that the process of local participation should go beyond merely making local communities the beneficiaries of marine ecotourism, elevating them to the status of equal partners in all of the key decision-making processes. Unless local people feel that they own the process of developing marine ecotourism in their area, they are unlikely to buy into it to the extent necessary to make it work in practice in the long term. A good example of this is the marine ecotourism marketing initiative currently taking place in West Clare, Ireland. Here, members of the local community were encouraged to participate in every stage of the project. This included developing an ecotourism brand for the area, developing a marine ecotourism website and marketing literature for the promotion of marine ecotourism in their area and developing codes of best practice for local ecotourism operators.

A common stumbling block for developing meaningful local participation in planning, managing and developing marine ecotourism is, however, that local communities are typically unused to participating in this way. Indeed, this book suggests that local people's attitude to participation is a problem for marine ecotourism projects as geographically dispersed as the island of Apo in the Philippines, County Clare in the West of Ireland, and the Avoca Beach Rock Platform in Australia. The problem is that local communities tend to be unaccustomed to engaging in a process of participatory development, their past experience typically having been one of outside 'experts' being drafted in to solve their problems for them. Consequently, there is a very real need for capacity building among local communities involved in marine ecotourism, enabling them to develop the skills necessary for effective participation. The use of educational programmes, not only to develop local people's awareness and understanding of the natural area concerned and environmental issues in general, but also about the nature of tourism and marine ecotourism more specifically, may represent an important first step in this process.

Another important reason why local participation is essential to genuine marine ecotourism is that the social element of sustainability requires that the impacts of marine ecotourism on the local community be taken into account: negative impacts will need to be addressed and positive impacts identified and built upon. This process will inevitably require substantial amounts of information, not only baseline data on various aspects of the social fabric of the local community, but also monitoring data on the local community impacts on marine ecotourism over time. Local participation is an effective means of gathering and interpreting such data. Local people inevitably know more about the condition of the local community and the impacts that marine ecotourism is having on it than park managers or project management staff, the latter often comprising people who are drawn from outside of the local area. Local people are also inevitably better at identifying relevant issues in the local context and they tend to be better informed about what the key constraints might be and who are the key players in the local community.

Encouraging local participation may also pay dividends in terms of managing the impacts of ecotourism to ensure that it takes place in as sustainable a manner as possible. Local people often have local knowledge about the environments in which ecotourism takes place, and may have a folk knowledge of how the natural environment has responded to pressures exerted on it in the past. This knowledge can be of vital importance if those involved in marine ecotourism are to establish and maintain a sustainable relationship with the natural environment in which it takes place. Local knowledge can be particularly important in situations where formal scientific research into the target species of marine ecotourism and ecosystems in which takes place is still sparse, which is often the case in respect of marine ecotourism.

Environmental Protection as a Priority

Environmental protection must be a key priority of marine ecotourism if it is to be genuinely sustainable. The activity of marine ecotourism depends directly on the continued availability of a high-quality marine environment in which to operate. Ultimately, it is the marine environment that draws ecotourists and provides them with experiences for which they are willing to pay. If the marine environment becomes damaged or degraded, ecotourists may no longer wish to visit the location in question, seeking more satisfying experiences elsewhere instead.

It may be argued that the ecotourism-generated impacts on the natural resource base are often themselves relatively insignificant in comparison to the myriad impacts originating from outside the sector. These threats include, to list but a few, the destruction of wetlands, the dumping of sewage at sea, over-fishing, nutrient pollution from aquaculture, dynamite fishing in coral seas, the introduction of exotic species and transport-related global warming. Yet, for a number of reasons, the marine ecotourism industry must be serious about addressing its own impacts. First, as this book has already noted, marine ecotourism is clearly a rapidly growing activity, and given the increasing exposure of this sort of holiday activity in the media, this high rate of growth is likely to continue for the foreseeable future. In the case of scuba diving, for example, increasing numbers of people are taking basic certificates, age restrictions having recently been relaxed and cost no longer being a limiting factor for many people. New technologies and practices such as snuba (a cross between scuba and snorkelling) and reef walking have also been developed, allowing people access to fragile ecosystems without the need for such certification. Dive operators have also promoted 'try dive' experiences, which allow people to dive on reefs with only a couple of hours of basic training, thereby increasing accessibility even further. Marine ecotourism will, therefore, undoubtedly have a significant scale effect to contend with in the near future. Even impacts that appear minuscule at the individual level may be highly significant when multiplied by the thousands or even millions of ecotourists that may visit a particular location.

A second reason why the marine ecotourism industry may be well advised to address its own impacts is that in doing so it may be able to raise awareness about the fragile and interdependent nature of its resource base. It may also be able to educate operators, local communities and ecotourists about what needs to be done to protect the natural environment on which marine ecotourism depends. Indeed, education and interpretation on the part of ecotourism providers, if it is well done, can have a positive long-term influence on the adoption of environmentally responsible forms of behaviour on the part of the individuals receiving it.

A third reason why marine ecotourism should address its own impacts is to ensure that the claims it makes about itself are consistent, both with one another and over time. Marine ecotourism, and indeed ecotourism in general, has established itself as a paragon of environmental virtue. As such, marine ecotourism has been accorded a reputation – in many cases still unproven – to which operators must live up. If proponents of marine ecotourism fail to address the impacts of their activities, it is hard to see how other economic activities, with perhaps much greater relative impacts on the environment, can be expected to mend their ways. Furthermore, as mentioned at several points in this book, there is also the issue of distinguishing between genuine and counterfeit instances of marine ecotourism. If would-be proponents of genuine marine ecotourism do not keep their own house in order, it will be difficult, if not impossible, for the potential ecotourist to distinguish the good from the bad.

Another important issue discussed in this book is the need for adequate research into the various aspects of marine ecotourism that contribute to its sustainability or otherwise. Given that scientific knowledge about many ecosystems and particular species of wildlife that are the subject of marine ecotourism is still relatively sparse, a vital prerequisite for genuinely sustainable marine ecotourism will be to ensure that sufficient funds are generated for research. This needs to be fed from both from the natural sciences and the social sciences. Often, the primary rationale for developing marine ecotourism is, of course, precisely to enable research and to fund associated conservation efforts. This is particularly the case when marine ecotourism products are supplied directly by a protected area authority or a research organisation. In many other instances, however, private individuals or firms will operate the marine ecotourism activities in question. In such cases it may be considered justifiable to impose a levy on ecotourism operators to help pay for the necessary research and conservation work. A proportion of this levy may, of course, be passed on to the consumer (although not all of it, since the demand for ecotourism experiences is never going to be perfectly inelastic). While this outcome might at first seem to be undesirable, a moment's reflection confirms that the practice accords fully with the widely-accepted 'polluter pays principle', which states that those responsible for causing damage to the natural environment should be the ones to pay for such damage to be remedied.

In the absence of sound scientific knowledge about the potential environmental impacts of marine ecotourism and the tolerances of the marine environment to such activities, it would seem prudent to adopt the 'precautionary principle' in planning, managing and regulating marine ecotourism activities. According to the precautionary principle, activities should be restricted to levels of intensity consistent with scientific knowledge about their likely impacts on the natural environment in which they take place. This is the approach taken, for example, by the conservation plan adopted by the dolphin-watching industry in the Shannon estuary. Under the plan, the total time dolphin-watching vessels are permitted to be in close proximity with the dolphins was set at 200 hours for the years 2000 and 2001. A conscious decision was taken not to increase this unless the data provided by the dolphin-watching vessels showed that there would be no negative effect on the behaviour or habitat of the dolphins.

Balance of Statutory and Voluntary Approaches to Regulation

The regulation of ecotourism in terrestrial settings has tended to be based on a wide range of underlying approaches: from legislative measures, on the one hand, to voluntary codes of conduct, on the other. The same is largely true in the context of marine ecotourism, although there has been considerable emphasis on the more informal, voluntary measures. The reasons for this are not difficult to appreciate. Indeed, this book has already identified some of the major problems involved in the formal regulation of marine areas, including the open and dynamic nature of the marine environment, problems of spatial definition and jurisdictional coverage, and multiple-use conflicts. Such conditions make it difficult, if not impossible, for regulations relating to marine ecotourism to be designed and implemented effectively. In many cases, therefore, voluntary measures remain the most readily applicable means of regulating marine ecotourism.

The voluntary approach is not, however, without its inherent weaknesses. Where state bodies currently take responsibility for the management of a marine area, they are often unwilling to release control to the industry because they are afraid that they will lose the ability to influence the issues that are being regulated for. Indeed, a view that is commonly expressed is that voluntary measures often represent a practical first step toward the formal regulation of marine ecotourism activities. The Whale and Dolphin Conservation Society, for example, provides a generic code of conduct, designed to be adapted for local implementation, which is intended to fill the regulatory void until formal regulation is possible. In a similar vein, the International Whaling Commission has also suggested that voluntary guidelines or codes of conduct tend to lack effectiveness, and have called for the introduction of statutory regulations on whalewatching.

Many observers, therefore, recommend a judicious blend of statutory and voluntary measures. If this is done successfully, those responsible for the planning and management of marine ecotourism can receive the 'best of both worlds': the flexibility of the voluntary approach along with the 'safety net' of the statutory approach. This, of course, requires that the measures introduced are well designed, both in themselves and in respect of any associated measures. For example, voluntary measures must be clear and unequivocal: one of the main reasons they have often lacked effectiveness in the past is that it has not been clear to all concerned exactly what is required for compliance. Making voluntary measures simple and clear-cut may also help to negate the potential for counterfeit ecotourism operators to use spurious compliance with them as a marketing tool, another common problem in practice. One way of ensuring that codes of conduct are as effective as possible is to ensure that those who are going to use them have a substantial input in their development. Indeed, several cases have already been noted in this book where engaging in a process of consultation has proven invaluable in designing voluntary measures. The International Ecotourism Society's 'Marine Ecotourism Guidelines', for example, were not only based on the Society's two established codes of conduct (for nature tourism operators and for ecolodges), thereby allowing for lessons from previous work to be built upon, but were also informed by three different consultative exercises: a mail-based survey of marine ecotourism operators, a series of stakeholder meetings and final review by a panel of international experts.

The regulation of dolphin-watching in the Shannon estuary, Ireland, demonstrates how a judicious blend of voluntary and formal approaches might be achieved. A voluntary code of conduct has been introduced that specifies how both commercial dolphin-watching vessels and recreational craft operating in the estuary are expected to behave in the presence of dolphins. For commercial dolphin-watching vessels, this code of conduct is backed up by the status of the Shannon estuary as a candidate Special Area of Conservation (SAC). This status requires the owners of commercial dolphin-watching vessels to notify the management authorities of their intention to operate in the SAC and to seek written permission to do so. This is contingent on the agreement of the owner of the dolphin-watching vessels to abide by the code of conduct and conservation plan, to provide monitoring data and to show competence in environmental education and species identification. This, in turn, is reinforced by a purely voluntary measure, which is for accredited vessels to be issued with a flag that they can fly from their vessels and use as a kind of ecolabel.

Education and Interpretation

It is generally agreed that education and interpretation are fundamental elements of the principles of marine ecotourism. Indeed, the National Ecotourism Strategy in Australia places them at the very centre of both the concept and practice of ecotourism. This book identifies a number of important reasons why education and interpretation should be seen as key components of genuine marine ecotourism. The first is that if marine ecotourism aims to be a 'responsible' form of tourism, there needs to be some mechanism for getting the message of responsibility across to tourists. If this is not accomplished, tourists will simply not know how to behave in ways that support, rather than detract from, the conservation of the environments involved. While it might be possible for marine ecotourism operators to behave more responsibly in respect of their own activities (such as purchasing environmentally-friendly supplies or undertaking energy-efficiency measures in their offices), a significant part of the impact of their operations is likely to be imposed by their clients, the tourists themselves. This requires tourists, who will typically be uninitiated in such matters, to know how to conduct themselves in the course of their visit.

A second reason why education and interpretation might be considered central components of genuine marine ecotourism is that they can help add value to the ecotourism experience, increasing visitors' willingness to pay and enhancing the financial viability of ecotourism operations. Some marine ecotourism is provided by organisations whose primary purpose is to conduct research, with revenues from ecotourists helping to fund the costs of research (for example operating a research vessel). In this case, the role of education and interpretation in helping to achieve the wider objectives of the marine ecotourism organisation may be explicit and direct. In other cases, for example where operators pay a fee levied by a park management authority, the link may be indirect. The link is, however, equally important. As the example of the introduction of visitor fees to Bunaken Marine Park in Indonesia illustrates, tourists are usually more willing to pay access fees when they know that the funds will be used to help support the research and conservation activities of the park. Education and interpretation may also help to ensure that the benefits of marine ecotourism flow to the local community, particularly when local guides are used.

Education and interpretation may also be significant to marine ecotourism operations in helping them to manage people's expectations of the experience that is on offer. Often, target species are only relatively rarely observed in areas that are easily accessible to day-trippers. Basking sharks in the waters of South West England are a good example (see Speedie, this volume). Unpredictable and uncontrollable factors such as weather conditions may also prevent whale-watchers from viewing as many animals as they might have hoped. Education and interpretation are vital tools for managing people's expectations of how often and how intensely the interactions with the wildlife might occur, helping to ensure that they do not go away disappointed when sightings are few and interactions are brief. Unfortunately, the marketing activities of 'counterfeit' ecotourism operators that are using the title of ecotourism but not applying the principles, often 'hype' the experience, leading people to believe that encounters are more certain, regular and in-depth than genuine marine ecotourism operators can hope to achieve. A third reason for education and interpretation in marine ecotourism is, therefore, to help ensure that tourists' expectations are compatible with the kinds of experience genuine marine ecotourism operations are actually able to provide.

A fourth reason why education and interpretation might be seen as important in meeting the objectives of genuine marine ecotourism is that such activities enable wider environmental messages to be delivered. In effect, marine ecotourism can act as a vehicle for more general environmental education: for example to get across messages relating to issues such as recycling, water pollution, energy use and global warming. Marine ecotourism certainly has the potential to inspire people, helping them to understand the importance of their relationship with the natural world. If this inspiration can be turned into tangible actions when the ecotourist returns home, then ecotourism can serve a very useful social role in extending environmental good practice. Ultimately, this is also in the interests of marine ecotourism, since the environmental problems covered clearly represent impacts on the marine resources on which marine ecotourism depends for its sustainability. Of course, marine ecotourism also has a responsibility to minimise its own contribution to these wider environmental problems.

A major difficulty in making education and interpretation effective in the roles identified above is that marine ecotourism is, after all, just a form of tourism. Tourists are normally looking for a range of outcomes as they undertake their ecotourism experience. Having fun, relaxation, being with one's family and freedom from the conventions and constraints of everyday life are likely to be important motivations for the tourist. Being educated may be very low on their list of priorities, if indeed it is present at all. As a result, the marine ecotourism provider may find it difficult to get the message of responsibility across to their client using educational and interpretive techniques, however well designed and well implemented these might be. However, a major advantage of conducting education and interpretation in the marine ecotourism context is that the audience is often truly captive, being located aboard a vessel taking them to, from or around the geographical area in which target species or other attractions are to be found.

Collaborative Approach

Collaboration in the context of marine ecotourism can be either intra-sectoral (among operators functionally located within the marine ecotourism sector) or inter-sectoral (between operators within the marine ecotourism sector and operators in other tourism industry sectors). Marine ecotourism operators are, of course, intimately connected to one another at the local level because they share a common resource: the marine environment in which they all operate. Consequently, any negative impacts of irresponsible behaviour on the part of any single operator are likely to be experienced by all of the others. If one 'rogue' operator acts irresponsibly to damage or destroy the environment in which marine ecotourism takes place, all operators potentially stand to lose, the quality of experience they are able to provide being forced downwards. In some cases, where the behaviour of the target species is highly migratory, collaboration among marine ecotourism sector operators will need to take place at the regional, national or even international level.

The interdependence of the sea-based and land-based components of marine ecotourism also makes a collaborative approach desirable. The sea-based elements of marine ecotourism are often heavily dependent on sea conditions and therefore good weather. Even the hardiest marine ecotourist is likely to be put off by rough seas and poor visibility. Safety issues may also preclude the operator putting to sea at certain times. Land-based facilities may, therefore, serve as a useful complement to sea-based marine ecotourism. In many areas, for example, whales and dolphins can be observed from the land as well as from vessels at sea. Visitor centres, natural history museums and marine aquariums may also provide viable distractions for marine ecotourists who are unable to put to sea for various reasons. Land-based facilities can also help destinations to extend their tourism season, making ecotourism-based holidays in the area an attractive proposition even at the 'shoulder' periods of the year when sea conditions are not as dependable. Providing alternative, shore-based activities may also help to increase the average length of stay of visitors to the ecotourism destination. Collaboration can, therefore, represent a good strategy for making marine ecotourism a more viable proposition. Simply put, if such activities are integrated, their combined drawing-power can be greatly enhanced.

The economic benefits of marine ecotourism are also more likely to be maximised for a local destination if ecotourists sleep, eat and drink, and buy their equipment and souvenirs locally. If marine ecotourists are only day visitors – perhaps because accommodation in the local area is limited or of poor quality – the local economy will fail to reap the full potential of marine ecotourism to generate benefits. Marine ecotourism operators will, therefore, need to collaborate closely with the other elements of the local tourism industry if marine ecotourism is to achieve its maximum potential as a source of local economic benefits.

Of course, collaboration does have its problems, one of which is the uneven power relations that often exist among local stakeholders. Indeed, this book raises a particular concern about the degree to which collaboration in the context of marine ecotourism can truly serve to meet the needs and desires of the already marginalised segments of the local community. Typically, those lacking in education, wealth and social status (for example, women in many communities) are those least likely to benefit from collaborative approaches to developing, planning and managing marine ecotourism. This is somewhat ironic, in that while these are often the groups that marine ecotourism can help the most, they are often the least able to engage in marine ecotourism in a truly collaborative manner. The problem is especially acute in the marine ecotourism context, because the capital requirements for establishing a marine ecotourism business are typically higher than for land-based ecotourism. Marine ecotourism also tends to require relatively high levels of investment in craft, equipment and skills, implying that the task of enabling poorer people to benefit from marine ecotourism will require genuine effort and long-term commitment on the part of the government or protected area authority involved.

The Future

This book has demonstrated that marine ecotourism has considerable potential to serve as a more sustainable (or perhaps, more correctly, less unsustainable) form of tourism. Marine ecotourism may represent an effective means of regenerating peripheral areas in economic terms, by virtue of its potentially substantial multiplier effects and its ability to re-deploy existing infrastructure and skills. It may also represent a viable means by which local communities may be regenerated in social and cultural terms, breathing new life into areas that have, for one reason or another, lost their links to the sea. Marine ecotourism can also generate funds for researching, conserving and protecting the marine ecosystems and species that form our common heritage. At the same time, however, this book has also highlighted a number of important challenges to marine ecotourism attaining and maintaining this position. By raising the relevant issues and discussing real-world experiences of developing, planning and managing marine ecotourism, it is hoped that this book will serve to move this agenda forward.

One issue that has not been discussed in great detail in this book is that of the need for marine ecotourism to be marketed responsibly. Undoubtedly, however, this will be an increasingly important issue. Marketing is one of the means by which marine ecotourism is implemented in practice, since without marketing their activities marine ecotourism operators cannot hope to build up their customer base. Yet marketing that is inconsistent with the wider principles of ecotourism can compromise even the most carefully planned and managed marine ecotourism. Irresponsible marketing may, for example, encourage tourists to believe that they are going to get very close to the cetaceans they are viewing as part of a commercial whalewatching trip. In many cases this may be an unrealistic claim, particularly when the animals are not used to the presence of humans. Tourists who have their expectations 'hyped up' by irresponsible marketing are less likely to feel satisfied by their experience, and hence less likely to recommend the experience to others or to make return visits themselves. What is perhaps worse, however, is where such marketing encourages tourists to behave irresponsibly, for example by trying to touch whales or dolphins. Similarly, unrealistic marketing may encourage operators of so-called 'ecotourism' experiences to behave in irresponsible ways, for example by splitting mothers from their calves in order to gain a 'close up' whale encounter for their clients. As such, marketing and management are inextricably linked to one another in the context of marine ecotourism. In the future, management efforts will need to pay much closer attention not only to the existing resource-based and planning policies relating to use of the marine environment (the supply side) but also to the associated patterns of actual and potential demand for marine ecotourism experiences (the demand side). The marketing of marine ecotourism is clearly a very important link between these two sides of the equation.

A second important issue, which has only really been touched on in this book, is that of monitoring. Assessing whether or not marine ecotourism is being effectively planned and managed requires monitoring not only of the status of the marine environment and the behaviour of the animals targeted by marine ecotourism but also of the status of the economic and socio-cultural environments in which marine ecotourism takes place. Monitoring must also address the degree of compliance with regulations, on the part of marine ecotourism operators. This applies particularly to the long-term impacts of marine ecotourism. Marine ecotourism may well be effective in achieving a sustainable short-term, day-to-day relationship with the natural and socio-economic environments in which it is based but the possible longer-term, cumulative impacts of marine ecotourism must also be considered. The latter remains a considerable challenge for marine ecotourism, particularly in view of the inherent characteristics of marine environments that tend to frustrate attempts at long-term monitoring.

The final word in this book is, quite rightly we feel, one of exhortation to all those involved in marine ecotourism. There is no reason why marine ecotourism should be intrinsically sustainable. Ideally, the various activities involved need to be planned and managed in such as way that sustainability is built in from the outset. However, this is not to preclude established marine tourism operations from taking the decision to revise their practices and begin to work towards genuinely sustainable marine ecotourism. In any case, marine ecotourism that is not genuinely sustainable runs the risk of destroying the resource base upon which it depends, ultimately to the detriment of everyone. Yet marine ecotourism that is properly planned and managed according to the principles of sustainability can serve to support local economies, provide secure livelihoods for local people, regenerate coastal communities, and conserve and protect the marine environment for the benefit of both present and future generations. The key is to make sustainability the watchword for everything that goes on in the name of marine ecotourism. Only then can marine ecotourism hope to achieve the laudable objectives that are set for it. Making marine ecotourism truly sustainable is clearly not going to be an easy task but the experiences illustrated in this book at least give us some hope that success is possible.

Index